Theory of Innovation

Jati Sengupta

Theory of Innovation

A New Paradigm of Growth

Springer

Jati Sengupta
Department of Economics
University of California, Santa Barbara
College of Letters & Science
Santa Barbara, CA, USA

ISBN 978-3-319-02182-9 ISBN 978-3-319-02183-6 (eBook)
DOI 10.1007/978-3-319-02183-6
Springer Cham Heidelberg New York Dordrecht London

Library of Congress Control Number: 2013948467

Printed on acid-free paper

Springer is part of Springer Science+Business Media (www.springer.com)

That thou art

Once on a full moon night I saw the moonlight dancing in the rain.
All of a sudden He came out and stood in front full of smiles
And joy. Stop not, Go Forward: said Sri Ramakrishna, my Ishta,
my Ideal. May I follow him all my life.

Jati Sengupta

Preface

Modern economies today have undergone a dramatic change, thanks to the upsurge of computer and communication technology. Technology frontier today is driven by the information frontier and innovation provides the key catalytic driving force. Schumpeter views innovation as challenges: challenge to the Walrasian competitive equilibrium, challenge to the modern entrepreneurs, and challenge to management. Managing innovations in many forms and adopting forward-looking business strategies are important today for success in modern business enterprise. This new paradigm of industry growth and the impact of endogenous innovation provide the central focus of this volume. Technology diffusion, human capital deepening, dynamic efficiency, and market growth provide the key components of the modern theory of innovation. This theory has several basic features: (1) to explore a comprehensive theory of innovation extending the Schumpeterian perspective, (2) to develop a new theory of management that has been called the corporate lattice model, (3) to explore the need for collaborative ventures in R&D investment, (4) to discuss the many profound impacts of the Internet and associated technology, and (5) to explore the dynamic efficiency generated by the innovation frontier and its impact on economic growth under rivalrous competition.

Today's business leaders are aware that in this knowledge economy the quality of their workforce drives the value of their shares. According to a Brookings Institution study nearly 85 % of a company's assets are related to knowledge and talent. Because talent works at every level of the business corporation, the changes necessary to develop that talent extend to nearly every aspect of the company's activities. The shortage of critical talent now and in the near future is one big challenge for the managers today. The US Department of Education estimates that 60 % of all new jobs in the early twenty-first century will require skills that only 20 % of the current US workforce possess. Skill development and emphasis on innovative growth provide the key elements of successful management today. Need for effective collaboration is all the greater in this framework. Given the ever increasing pace of global business working together collaboratively becomes critical to keeping pace with innovation-intensive competition. Rather than focusing on defending a few key ideas or stocks of knowledge, companies must use the flows of knowledge generated by innovation to continuously accelerate newer and better ideas. This volume seeks to explore a comprehensive view of innovations in all its aspects. Schumpeterian models of innovation are extended in terms of modern theory and various challenges before modern management are discussed in some detail.

I would like to take this opportunity to express my deep appreciation to my wife who provided constant support and to my Guru for his encouragement. Both told me to remember that if even one student gets benefit from reading my book, I should continue to write it. My grandchildren—Jayen, Shiven, Aria, and Myra—helped me by always asking me what I am writing about. I had a hard time explaining to them, hoping that one day they would understand when they are grown up. May they lead an innovative life.

Santa Barbara, CA, USA Jati Sengupta

Contents

Chapter 1
Theory of Innovation

Innovation in a broad sense involves developing new processes, new products, or new organizational improvements for an industry. It can take many forms, but in every form, it tends to reduce unit costs and/or helps to expand market demand. Some of the important types of innovation are as follows:

1. Technology-based innovation
2. Endogenous v. exogenous innovation
3. Innovation in selection mechanism in industry growth
4. Innovation through technology consortium

Technology-based innovations include such forms as (a) product innovations, (b) industrial R&D investments, and (c) technology transfers through imitations and improvement. Endogenous innovation involves market incentives for developing the propensity to invest for innovations. The expected rewards for winning the technology race and the likely protection of monopoly products through patent laws provide the basic ingredients of such market incentives. Basic research in academic and nonprofit institutions provides an example of exogenous innovation. This type of R&D research provides the general background, which may sometimes lead to new products or new processes later on. Solow's growth model assumed all technological progress as exogenous, though eventually it led to an upward shift of the production frontier involving significant productivity gains for industry and the overall economy.

The next two types of innovations involve less tangible phenomena. The evolutionary selection mechanism chooses firms through dynamic market efficiency and the patent system. This mechanism recognizes both the cumulative nature of technological change and the endogenous aspects of market structures. The patent system is the second policy instrument of technological selection. It defines property rights which support the incentives for technological innovations. For example, the pharmaceutical industries undertake large R&D investment due to such incentives. In the technology literature it has been recognized that a patent system with a *limited scope* promotes technology and knowledge diffusion, which innovations with a *broad scope* encourage more experimentation in technological research.

J. Sengupta, *Theory of Innovation: A New Paradigm of Growth*,
DOI 10.1007/978-3-319-02183-6_1,

Innovations in human capital generally occur through learning by doing. Three types of measures of learning by doing are used in the econometric literature. One is the cumulative research experience embodied in cumulative output, where the latter is often taken as a proxy for technological progress, e.g., the empirical studies of industrial productivity growth by Norsworthy and Jang (1992) have found the cost of reducing effort of such technological progress to be substantial in microelectronics, telecommunications, and related industries. The second measure is cumulative experience embodied in specific strategic inputs like specialized capital. A third measure of experience in knowledge capital may be due to the imitation process, whereby the spillover of knowledge from technologically advanced firms to others is captured by the followers. Growth miracles in Southeast Asia in the last three decades have shown evidence of such innovations in knowledge transfer.

The impact of innovations on industry growth has significant economic implications for theory and experiences of economic growth of nations. It is useful to critically review this framework here.

A major impact of innovation is to reduce unit costs of production and distribution. Hence it reduces prices and thereby improves competitive advantage. Creative destruction is the process by which old sources of competitive advantage are destroyed and replaced with new ones. In Schumpeter's theory of innovation the innovative role of the entrepreneur is to exploit the shocks or discontinuities that destroy existing sources of advantage. Porter (1990) in his *The Competitive Advantage of Nations* views competition as an evolutionary process. Firms initially gain competitive advantages by altering the basis of static competition. They win not just by recognizing new markets or technologies but also by moving aggressively to exploit them. A firm's home nation plays a critical role in shaping managers' perceptions about the new opportunities that can be exploited. The domestic economic framework shaped by technical and scientific education helps put pressure on firms to innovate, invest, and improve R&D activity. Thus innovation increases comparative advantage of small countries in international trade that are rich in technological knowledge. As examples one may refer to countries like Taiwan and Finland. The world Economic Forum Report edited by Porter (2004) has computed a growth competitiveness index (GCI) based on three components: infrastructure development, quality of public institutions, and the adoption of the best practice technology of the world. Its report for 2002–2004 showed the following ranking:

Clearly Taiwan's record of performance is most impressive. We may note also that in terms of the average number of annual US patents per million people, the top rankings in the world in 2004 were (1) USA, (2) Japan, and (3) Taiwan. One of the

	Rank		Technology rank
	2002	2003	2003
Taiwan	6	5	3
Finland	1	1	2
Korea	25	18	6
USA	2	2	1
China	38	42	65

major forms of innovation involves R&D and the associated investment process in knowledge creation and diffusion. Research in "knowledge capital" captures the external economies of R&D done by other firms. Thus the external benefits of R&D investment in developed countries spill over to other developing economies. This provides one main reason of growth miracles in Southeast Asia over the last three decades. The R&D race provides for the winner quasi-monopoly profits in the short and medium term. This increases the market power and dominance of large firms. Schumpeter emphasized this aspect in his dynamic theory of evolution of firms under innovation. A dominant established firm's incentive to innovate may be weaker than that of a smaller firm or a potential entrant. The sunk cost and replacement effects weaken the established firm's incentive to innovate. The efficiency effects of R&D-based innovation strengthen the dominant firm's incentive to innovate compared to a potential entrant's incentive. The reason is that the incumbent can lose its monopoly, if it does not innovate, whereas the entrant will become at best a duopolist if it succeeds in innovative venture.

1.1 Technology and Efficiency

Innovation as technology is most important for its efficiency effect. The technology process comprises several stages. Pure research, oftentimes in academic and public institutions, provides the basics of applied research. In this general sense, knowledge may be viewed as capital, which provides the basis of a complementary input in the production function. Technology creation and diffusion help an economy to build new types of dynamic efficiency such as innovative efficiency and access efficiency. Innovative efficiency occurs through competitive advantage gained through new knowledge. Access efficiency begins through globalization of markets, where networking and scale economies in knowledge-intensive products such as computer hardware and software, telecommunications, and pharmaceuticals have intensified the innovation capabilities of modern firms. In modern times economies have undergone a dramatic transformation from large-scale material manufacturing to the design and use of new technologies, software innovations, and social networking like Facebook in telecommunications. These new innovations are all characterized by increasing returns and scale economies, and also these have positive feedback and strong complementarity effects through knowledge diffusion and transmission. There exist five main reasons for the recent upsurge of these new innovations: (1) high fixed costs with low variable costs so that the marginal cost is very small, (2) network effects by which the value of a product increases with the number of users, (3) high switching costs which imply that users tend to stay with the product of technology for a minimal time, (4) externalities of new processes diffused to other countries and other industries through spillover effect of R&D investments in knowledge capital, and, finally, (5) nanotechnology has spread the speed of miniaturization to various products and services with a complementary impact on various interrelated products and services.

Recently Nachum (2002) tested several hypotheses over US panel data for 1989–1998 comprising 650 firms in order to test the role of innovative activity on foreign direct investment by multinational firms. One of his significant findings is that the impact of innovative investment is much stronger for IR (increasing returns) dominated industries than for the DR (diminishing returns) industries. Also networking, entrepreneurship, and flexible organizational structure of the IR industries played a similar role.

It is interesting to note that the innovative role of modern IR industries was predicted by the Schumpeterian model of innovation. Schumpeter distinguished between five types of innovations as follows:

1. Product innovation, where a new type of product or service is added to the list of goods requiring a change in the production routine.
2. Process innovation, which entails a change in the production function or the production routine. This frequently involves a change in the input mix and input quantities.
3. Organizational innovation, which involves change in the managerial routines leading usually to a change in market structure, e.g., a reorganization of a price cartel.
4. Market innovation, where a product is introduced to new markets like selling abroad.
5. Input innovation, which involves new raw material or new intermediate good, e.g., new sources of energy or new types of uses of the existing inputs.

The central dynamic role in these innovations is played by the entrepreneur in Schumpeter's model. The Schumpeterian entrepreneur (S-entrepreneur) plays a dynamic leadership role as soon as a significant innovation is made in the previously stationary equilibrium economy. In his words the entrepreneur is the king, the banker, and the ephor of the market. He is the "king," because he has the will and energy to initiate the transformation of the system of routine. He initiates a selective pressure on the incumbent firms, who either go bankrupt, exit the market, or adapt to the change. The essence of the entrepreneurial function comprises economic activities of the following types:

1. The production of new products or services or new qualities of goods.
2. The introduction of new production technologies.
3. The creation of new forms of industrial organization at different levels of business, e.g., increasing use of venture capital or hedge fund investment in today's stock market.
4. The opening up of new markets, e.g., globalization and widespread use of networking methods in business communication and finance.
5. The opening up of new sources of supply through widening of the supply chain and global investment by multinational firms. The tide of significant economic growth in the newly industrialized countries (NICs) of Southeast Asia over the last three decades bears eloquent evidence of this openness process in international trade.

6. The diffusion of new innovation in different sectors and the spillover effects of modern technology alluded to by Schumpeter have been strongly emphasized in recent growth literature.

This list of economic activities of the innovating S-entrepreneur may be augmented by two recent developments. One is the concept of "hypercompetition" introduced first by D'Aveni (1994) and the other the concept of "evolutionary efficiency" studied in evolutionary biology and applied recently in evolutionary growth theory in economics. In software research and other high-tech fields of today's business, intense competitive pressure has generated four types of dynamic efficiency analyzed in some detail by D'Aveni. There are *production efficiency* in terms of a decline in unit costs, *innovation efficiency* in terms of R&D investment and race for patents, *access efficiency* where the innovating firm races up the escalation ladder and through mergers and buy-ups keeps out potential entrants, and *resource efficiency*, where the companies seek to expand their resource base through multinational world markets. Hypercompetitive firms must use their fixed assets and accumulated resources to build their next temporary base of competitive advantage. Thus IBM bet the company on the 360 series computer and the bet paid off in the 1960s through increased market dominance and large profits due to specific competitive advantage. But its resource base could not sustain this dominant position very long due to its failure to diversify. Small competitors like Apple and Microsoft became giants by seizing the new opportunities for developing PCs by their diversified resource base and its efficiency.

One has to note that a new innovating firm's ability to gain competitive advantage over others depends on the presence of "the market for research ideas" in the industry. Teece (1986) identifies two basic elements behind the market for ideas. One is that the technology is not easily expropriable by others. This may be due to the requirement of large fixed investment in R&D, e.g., new medicine or drugs. The second element is the existence of specialized assets in the company such as specific product capabilities or core competence that must be used in conjunction with the innovative product. The dynamic side of the competition is an evolutionary process, where the innovating firms gain competitive advantage by altering the P=MC basis of static competition. They win not just by recognizing new market or new technologies but also by moving aggressively to exploit them. They sustain their advantages by increased investment for improving the existing sources of advantages and for creating new ones. These advantages form the basis of the concept of "core competence" of a firm, which is so strongly emphasized by the management science experts. The traditional economists have failed to emphasize these managerial aspects of dynamic efficiency in the competitive framework.

1.2 Endogenous Aspects of Innovation

Endogeneity of industrial innovation has three basic sources:

1. The market structure and its impact on the development of new technology

2. Endogenous growth theory and the impact of capital accumulation
3. Knowledge diffusion and its impact on the spillover of new technology in globalization

The market structure basically embodies profitability and its long-run sustenance. The profit incentives guide the entrepreneurs along the path of innovation in various forms. Schumpeter emphasized the distinction between innovative and noninnovative agents or firms. This distinction allows the Schumpeterian entrepreneur (S-entrepreneur) to play a more dynamic role in industry growth than the Walrasian entrepreneur (W-entrepreneur). The W-entrepreneur is a core noninnovative agent who adapts promptly to a change of the economic system thus contributing to the equilibrium of the system. By contrast the S-entrepreneur disturbs the static competitive equilibrium by buying or using resources to change one or more parameters of the economic system. The S-entrepreneur wants to change what to others appears to be a given production routine and if necessary the related consumption routines. The S-entrepreneur can base the evaluation of the profitability of his project on the Schumpeterian conjecture that the other firms do not adapt quickly. This conjecture of Schumpeter, very similar to Cournot's conjecture, is in sharp contrast to the extremely flexible behavior of W-entrepreneurs. Andersen (2011) has argued that Schumpeter ultimately wanted to endogenize economic institutions, science and invention, and parts of behavioral psychology in his dynamic evolutionary process of innovation.

Recently innovations flow has been modeled as a stochastic process for the competitive R&D race. If we assume that successful innovations arise as a result of a Poisson stochastic process with an intensity u, then the probability of a firm innovating successfully during period dt is udt. The expected monopoly profit (π) for the successful S-entrepreneur may then be written as

$$\pi(n, u) = r(n)u - c(u, f), \tag{1.1}$$

where $r = r(n)$ is the instantaneous monopoly surplus (profit) obtained by the winner of the innovation race and $c(u, f)$ is the firm's cost function which is assumed to be convex on the intensity u and fixed cost f. Here the monopoly profits or surplus is assumed to depend on the number of firms in the industry. Fölster and Trofimov (1997) maximized the profit function (1.1) with respect to intensity u and the result is an optimal profit function which is S-shaped. This type of profit function implies that the positive effect of R&D innovations sometimes dominates the negative effect of increased competition. Spence (1984) viewed R&D investments basically as fixed costs, which reduce unit cost. In many instances, e.g., new medicines, the R&D expenditures take the form of developing new products at cheaper prices. In this environment market structures are likely to be concentrated and imperfectly competitive. What is significant about R&D innovations is not only product differentiation and scale economies but also the spillover effects of externality effects. The benefits of R&D spread to other firms through learning by doing and knowledge diffusion. Spence modeled this process in terms of the dependence of unit costs $c_i(t)$ of firm i on the accumulated knowledge $z_i(t)$, where

$$\dot{z}_i(t) = m_i(t) + \theta \sum_{j \neq i} m_j(t) \tag{1.2}$$

the dot denotes the time derivative and $m_i(t)$ the current expenditure of firm i on R&D. The parameter θ $(0 \leq \theta \leq 1)$ captures spillover effects where unit cost

$$c_i(t) = F(z_i(t)) \tag{1.3}$$

is a declining function of $z_i(t)$. In this model the case $\theta = 0$ represents no spillovers, while $\theta = 1$ represents the case when the benefits of each firm's R&D are shared completely. Spence derives an important relationship in this model relating the industry's total investment in R&D as a function of z as follows: R&D costs at the industry level

$$= zn/(1 + \theta(n - 1)). \tag{1.4}$$

This is a symmetric case with all firms alike. For a given level of z and $n > 1$, the R&D costs of the achieved amount of cost reduction decline as θ increases. As n tends to infinity, R&D costs tend to the upper limit of $1/\theta$ when θ is positive. For zero θ the R&D costs are proportional to the number of firms. Two implications are important. One is that the spillovers reduce the industry level costs of R&D for achieving a given level of cost reduction, though they may reduce the incentives for cost reduction. But the incentive reduction may be restored through appropriate policies of state subsidies. Secondly, when n decreases, the market becomes more concentrated. The incentive for temporary monopoly profit tends to be more dominant. The impact of ignoring spillovers is to make the investment decisions of firms more aggressive, because the anticipated return is perceived to be higher than it actually is. Due to this spillover effects, knowledge diffusions have intensified in recent years through software technology and increased direct investment by multinational corporations.

Recent developments in endogenous models of economic growth emphasized two key sources on endogeneity. One is knowledge creation associated with investment. A firm that increases its physical capital learns simultaneously how to produce more efficiently. This positive effect of experience on productivity is called "learning by doing," a term first coined by Arrow (1962). The second is the spillover effect from a firm to the industry and from industry to the overall economy. Consider for example the simplest endogenous growth model known as the AK model where output Y is

$$Y = AK, \tag{1.5}$$

where A is technology and K is capital including both physical and human capital. If A is assumed to be constant, so that there is no technological progress in the Solow sense, then one obtains in per capital terms $y = Ak,\ y = Y/L,\ k = K/L$. This implies that the growth rates of income y and capital k are equal

$$g_y = \Delta y/y = g_k = \Delta k/k = \dot{k}/k$$

and the savings investment equilibrium implies

$$g_y = \dot{y}/y = sA - (n + \delta) = g_k.$$

Thus the AK model can display positive long-run per capita growth without any technological progress, where technological progress is measured by a positive value of $\dot{A}/A$. Per capita income growth can occur even with zero technological progress. Note also that a higher savings rate s and a higher level of A can increase the long-run growth rate in this endogenous model of growth.

The key to endogenous growth in the AK model is the absence of diminishing returns in the factors that can be accumulated. Both Arrow (1962) and Romer (1990) attempted to eliminate the tendency for diminishing returns by assuming that knowledge creation was a side product of investment. A firm that increases its physical capital learns simultaneously how to produce more efficiently. This positive effect of experience on productivity is called learning by doing. Also each firm's knowledge is a public good that any other firm can access at negligible costs. In other words once discovered, a piece of knowledge spills over almost instantly across the whole economy. Lucas (1993) used this idea in explaining the Asian growth miracle, where countries like Taiwan, South Korea, Singapore, and China grew at a faster rate over the last three decades.

Empirical data seem to show that growth in knowledge capital, openness in trade, and foreign investment in these newly industrializing countries (NICs) of Southeast Asia have greatly contributed to their success rate. For example, South Korea's export growth rate of 22.9 % over the period 1965–1987 accompanied the average income growth rate of 6.4 %. China's reform of its national innovation system started in the 1990s. A good measure of R&D intensity is the ratio of R&D expenditure to GDP. By this measure China's R&D intensity rose from 0.74 in 1991 to 1.23 in 2003. Now in 2011 it exceeds 2.0. For Korea it rose from 1.92 to 2.96 during 1991–2004. Taiwan's contemporary knowledge-based economy has revealed remarkable growth of the information technology (IT) sector than China and other NICs of Asia. From 1995 to 1999 Taiwan's IT industry ranked third in the world after the USA and Japan. The overall R&D intensity rose from 1.78 in 1995 to 2.16 in 2003 and has exceeded 3.00 in 2011.

Recent empirical studies have shown that the R&D investment generating new industrial knowledge capital know-how in the USA has spread to the developing countries, and the successful NICs in Southeast Asia have taken full advantage of it. Many incremental innovations that come in small steps, e.g., in software and communication fields, have generated new innovations in these NICs and this is intensified by the boom in exports of technology-intensive products and processes.

It has to be noted that the knowledge diffusion process embodied in endogenous growth theory is rarely the dramatic breakthroughs that Solow and Schumpeter may have had in mind, but rather small improvements and small dispersal of a new process or product in which novelty and imitation imperceptibly get mixed up into one another. Two aspects of this incremental diffusion process deserve special mention. One is that technical knowledge is itself a kind of capital good embodied as K in the AK model. It can be stored over time because it does not get completely used up whenever it is put into the production process. The second important element is the impact of incremental innovation in terms of economies of scale and economies of scope associated with R&D investment. The innovation through R&D expenditures exceeds 5 % of total sales at many high-tech companies such as Intel, Microsoft, GlaxoSmithKline, and GE. The pharmaceutical companies spend upwards of $500 million to successfully develop a new drug. This contains a substantial indivisible investment, implying that every unit cost will decline very rapidly as the sales of the drug increase. Thus R&D expenditures also entail substantial economies of scale, since ideas developed in one research project create positive spillovers to another project. This happens more often in pharmaceutical and software firms.

1.3 Selection and Industry Evolution

Industry evolution depends basically on the selection mechanism operating through innovations and dynamics of the market structure. New technology and the process of entry and exit of firms are important forces here. Several factors play critical roles in the selection process. First, we have the evolutionary approach which emphasizes the firm's ability and competence to alter the market structure significantly. Following the Schumpeterian theory of technological innovations where size begets size and the cumulative processes of innovations generate significant scale economies. Here industrial dynamics would be characterized frequently by nonlinear and path-dependent processes, where random events like a new incremental innovation or a new software may have lasting and irreversible effects on the dynamic evolution of the selection process. Secondly, firms differ significantly in their commitment and ability to innovate. Thus innovations are largely endogenous to the firm through R&D investment and learning by doing. Thirdly, the evolutionary forces of selection which allow only some firms to survive and grow are subject to initial stochastic mechanisms which play an active dynamic role. Thus Jovanovic (1982) and Mazzucato (2000) have considered cost efficiency under innovation depending on a stochastic parameter. Uncertainty represented by the stochastic parameter is generally very high in the early stages of the industry life cycle, when the product design has not yet been standardized. In this phase the flexibility of small new firms allows them to be the leaders in cost reduction and new innovative experiments thus causing high rates of entry. During the mature stage of the industry life cycle, however, economics of scale and learning by doing favor large size firms and innovations become increasingly path dependent, thus yielding a more stable oligopolistic structure.

The entry process may be viewed in two broad ways. One is in terms of the entrant's output or price behavior vis-á-vis the incumbent. For example, if y_1 and y_2 are the two output shares of total industry output, then changes in y_1 may depend on the cost price strategy of the incumbent or the innovation efficiency (I_1) of the entrant, i.e.,

$$\dot{y_1} = dy_1/dt = f(\bar{c} - c_1, I_1),$$

where $c_1 \leq \bar{c}$ with $\bar{c}$ as the industry average unit cost. Secondly, the price (p) quantity (q) adjustment process may reflect the entry–exit dynamics as

$$\dot{q} = a\,[p - c(q)]\,;\; \dot{p} = b\,[D(p) - q]\,.$$

Here dot denotes the time derivative, $c(q)$ is unit cost, $D(p)$ is demand, and a, b are positive constants reflecting speed of adjustment. The competitive advantage principle underlies this growth mechanism which assumes a Walrasian framework.

1.4 Innovation Through Technology Consortium

The industry selection of successful innovating firms is motivated by the need to capture the spillover effects and the potential monopoly profits arising from potentially successful new innovations. In recent times intense competition in modern technology-intensive industries has forced firms to explore ways for sharing costs in large-scale R&D investment in knowledge capital. By pooling R&D resources the firms can exploit economies of scale and productivity gains. Such pooling of resources on the input side is not disallowed by the existing antitrust regulations in advanced industrial countries, although the pooling on the product side in the form of cartel or monopoly is not allowed. For many small- and medium-size companies the network sharing of resources with similar needs and demand structure has become critical to business success. Sharing communications through satellite systems (e.g., the *Cloud* in Internet systems) has become a common phenomenon. This type of sharing phenomena may often lead to consortiums or mergers involving interfirm cooperation at a horizontal level comprising similar firms. We discuss four types of such coalitions or consortiums:

1. Improving production efficiency through coalitions
2. Cost sharing of fixed capital investments
3. Sharing technology to internalize the spillover effects
4. Sharing of innovation costs through discriminatory pricing.

The static efficiency of competitive equilibria can be analyzed most easily through a Pareto efficiency model. Recently the method of data envelopment analysis (DEA) has been applied for analyzing Pareto efficiency. We analyze here two types of sharing by employing the DEA method. One uses the game-theoretic concept of efficiency and the other considers cost sharing of fixed capital investment in the common network system, for example. Such a strategy reduces each member firm's variable cost of production, because the cost of pooled investment in the network is usually subject to subadditivity, i.e., increasing returns to scale.

For Pareto efficiency in the DEA framework consider for simplicity one input (y_j) and m inputs (x_{ij}) for each firm j $(j = 1, 2, \ldots, n)$ and the reference firm k is tested for Pareto efficiency in terms of the following LP model:

$$
\begin{aligned}
&\min_{\beta} c_k = \sum_{i=1}^{m} \beta_i x_{ik} \\
&\text{subject to (s.t.)} \\
&\sum_{i=1}^{m} \beta_i x_{ij} \geq y_j \\
&\beta_i \geq 0; i = 1, 2, \ldots, m; j = 1, 2, \ldots, n
\end{aligned}
\tag{1.6}
$$

of the optimal values β_i^* such that if $\sum_{i=1} \beta_i x_{ik} = y_j$ and all slack variables are zero, then the firm k is Pareto efficient. If however $\sum_{i=1} \beta_i x_{ik} > y_j$, then it is Pareto inefficient, since the optimal output y_k^* defined by $\sum_{i=1} \beta_i x_{ik}$ exceeds the observed output y_k. Now consider the Pareto efficiency model (1.6) as a production game, where the n firms are n players. They can form a coalition of different sizes. Let $N = \{1, 2, \ldots, n\}$ denote the grand coalition of all n players and let S be a proper subset of N. Consider the dual of the LP model (1.6) as

$$
\begin{aligned}
&\max z = \sum_{j \in S} y_j \lambda_j \\
&\text{s.t.} \sum_{j \in S} x_{ij} \lambda_j \leq \hat{x}_i, i = 1, 2, \ldots, m \\
&\lambda_j \geq 0; j \in S,
\end{aligned}
$$

where $\hat{x}_i = \hat{x}_i(S)$ is a particular allocation of input i from the total inputs $\sum_{j \in S} x_{ij}$ held by the coalition. We may now form the characteristic function $v(S)$ for any coalition S as

$$\begin{aligned} v(S) &= \max \sum_{j \in S} y_j \lambda_j \\ \text{s.t.} \sum_{j \in S} x_{ij} \lambda_j &\le \hat{x}_i(S);\ \lambda_j \ge 0;\ j \in S. \end{aligned} \tag{1.7}$$

In terms of the cost-oriented Pareto model (1.6) we could specify a similar model for a fixed-size coalition S_r $(1 < r < N)$:

$$\begin{aligned} \min_{\beta} c_k &= \hat{x}_k(S_r)'\beta \\ \text{s.t.} \sum_i \beta_i x_{ij} &\ge y_j;\ j \in S_r;\ \beta_i \ge 0 \\ \hat{x}_k(S_r) &= \hat{x_{ik}}(S_r). \end{aligned} \tag{1.8}$$

Here prime denotes transpose and β is the column vector with elements β_i.

Several incentives exist why the competing firms bid for higher input allocations $\hat{x}_k$. One incentive is that by joining a particular coalition, a firm can reduce its unit costs by more than the case of not joining. Phlips (1995) has discussed this problem in some detail in order to show that explicit collusion should be analyzed as the solution of a noncooperative game. A cartel will appear as a noncooperative Nash equilibrium: each cartel member decided to join the cartel because joining is his best reply strategy. In particular to accept a production quota is a best reply strategy. From a policy point of view an interesting question is: in what sort of industries is explicit collusion likely to occur? It is generally held that cartels are more likely in concentrated industries where competitors are few. But this is a very vague statement. Noncooperative game theory gives a more clear-cut answer, as shown by Selten (1973). He showed in his paper entitled "A simple model of imperfect competition where four are few and six are many" that if there are less than five competitors, they will *all* find it profitable to enter into an explicit collusion, but if there are more than five competitors, it becomes more advantageous to stay out of cartels formed by others.

A second incentive for forming or joining coalitions is that some inputs like R&D investment through which most of modern innovations occur possess significant economies of scale and economies of scope, which may lower unit costs further and thereby increase profits. Since the industry decides on the optimal number of firms surviving by minimizing the total cost of producing the aggregate output given by market demand, the innovation process may squeeze out relatively inefficient firms from the industry.

A third incentive for collusion is to capture a large part of the spillover effects or externality benefits from modern day innovations in such fields as software, miniature engineering, and knowledge innovations in communication fields.

Finally, the input allocation game defined by the input-oriented Pareto efficiency model (1.8) may be used to characterize a nonempty core as defined in game theory. For instance consider the following three types of coalitions in terms of the model (1.8):

1. One player coalitions: $S = S_1$
   Here the n players act independently of each other and the minimax equilibrium concept can be employed as one possible solution. Thus, player k assumes all others as rivals, who select their strategies so as to maximize his cost c_k defined in (1.8). Let $\beta(N-1)$ denote the $(n-1)$-tuple vector with elements $\{\beta(1), \beta(2), \ldots, \beta(k-1), \beta(k+1), \ldots, \beta(n-1)\}$ and β_{N-1} be the feasible set of admissible strategies. Then $\beta^*(k)$ is a minimax strategy for player k, if and only if for all $\beta(k) \in B_k$ we have

$$\min_{\beta(N-1)} c_k\left(\beta^{**}(k), \beta(N-1)\right) \leq \min_{\beta(N-1)} c_k\left(\beta(k), \beta(N-1)\right),$$

   where $\beta^{**} = \max c_k$. This type of solution first selects the worst-case scenario for player k. Then it chooses the best of the worst.
2. n-player Coalition
   Here each of the n players agrees to form a cartel to achieve a lower cost than the minimax case.
3. r-player Coalitions $(S_r : 1 < r < n)$
   Here we have a coalition of r members. The concept of Pareto optimality is useful here. Thus the vector $\beta^*(R) = \{\beta^*(1), \beta^*(2), \ldots, \beta^*(r)\}$ is Pareto optimal up to coalition of size r, if and only if we have

$$\begin{aligned}\Delta c_k = &\max_{\beta(N-R)} c_k\left(\beta(R), \beta(N-R)\right) - c_k\left(\beta^*(R), \beta(N-R)\right)\\ &\geq 0 \forall k \in (1, 2, \ldots, r)\end{aligned}$$

   and $\Delta c_k > 0$ for at least one k.

Now let P be the set of all possible coalitions S. Let $\bar{\beta^*}$ be optimal for each such coalition in P (i.e., collective optimality), the vector $\left\{\bar{c_1}(\bar{\beta^*}), \ldots, c_r(\bar{\beta^*})\right\}$ belongs to what is known as the *core* of the game. Three implications of the core concept may be noted here. First, consider two coalitions—S and $N-S$—where N is the grand coalitions. Let $v(S)$ and $v(N-S)$ be the characteristic functions or values as defined before. If the coalition S is aware that the grand coalition has positive surplus defined by $[v(N) - v(S) - v(N-S)]$, then one can define the aggressiveness of coalition S as the share θ_S, where the claim of coalition S would be

$$R(S) = v(S) + \theta_S\left[v(N) - v(S) - v(N-S)\right].$$

With $\theta_S = 0$ for all coalitions we have the ordinary core and $\theta_S = 1$ implies the maximally aggressive core. The case $0 < \theta_S < 1$ defines the intermediate degree of aggressiveness of the core.

Secondly, let Q be any of three coalitions S, N, and $N - S$ and consider the DEA model (1.7) with S replaced by Q. We can then compute the value of $R(S)$ above for various values of θ_S that denote different rules of sharing the surplus. Thus in the race for innovations in R&D the potentially successful firms may stake for a higher share of the surplus.

Finally, we may view the DEA model before in two stages. In the first stage we determine the subset $n_{(1)}$ of n firms which are technically efficient in the sense of Pareto optimality. In the second stage we consider an allocation game among the $n_{(1)}$ efficient firms. The industry competition then selects the best of these firms by minimizing the total costs $C = \sum_{j=1}^{n_{(1)}} c_j(y_j)$ subject to the condition that total supply $(\sum y_j)$ equals total market demand. Here $c_j(y_j)$ is the cost function of firm j and if it is assumed to be quadratic for instance, then the efficient allocation among the active firms with positive outputs makes them all have the same marginal cost in a competitive framework. Under monopoly or collusion a conscious mechanism secures economic efficiency, because the cartel actively seeks a maximum net return.

Sharing of fixed capital investment among firms is becoming increasingly important in modern high-tech industries today like computers and telecommunications. Also biomedical and pharmaceutical research involve very long gestation periods exceeding 6–8 years and whenever successful they involve patents in order to protect monopoly rights for 15–20 years. Both of these investments have significant economies of scale, both internal and external. Internally it means that a firm's unit costs decline as the size of the total investment increases. Thus by pooling such fixed investment the industry can lower its average cost of production. Externally the economies refer to the fact that the expansion of total industry output lowers the total cost curve of most of the firms in the industry by dint of sharing the free benefits of total knowledge capital and know-how in the whole industry without paying any cost separately. In terms of actual economic policy the commission of the European Communities (EC) and judgments made by the European Court of Justice have make EC competition rules for private undertakings so that sharing of fixed investment for R&D and related knowledge capital is actively promoted in most of the high-tech fields. Articles 85 and 86 of the EC Treaty (the "Treaty of Rome") detail these negotiations. For example, Article 85, paragraph 1 prohibits (a) all agreements between undertakings and (b) all "concerned practices" which may affect trade between Member States and which have as their object of prevention, restriction, or distortion of competition. The agreements under (a) include price-fixing agreements, market-sharing agreements, quota cartels, explicit collusion on the product side, and so on. The "concerned practices" under (b) comprise "tacit collusion" which leads to "collusive outcomes" without there being explicit cooperation between the colluders. It has to be noted that exchanges of information between competitors on market conditions and on prices, firm specific production or export deliveries, and the like are considered by the EC authority as proof of tacit collusion. Van Bael and Bellis (1990) have analyzed in some detail the competition rules developed in the EC and the European Court of Justice. They have noted that if a collusive outcome is reached by noncooperative behavior of competing firms, then there is no collusion in a legal sense in the USA. This is a fundamental difference between competition law in the USA and the EC, as emphasized by Phlips (1995) in some detail.

Article 85 of the EC Treaty contains exceptions in paragraph 3 as follows: The provisions of paragraph 1 may be declared inapplicable in the case of agreements and concerned practices which contribute to improving production or distribution of goods or to promoting technical or economic progress without imposing unnecessary restrictions. One such important exemption was granted for the creation of joint ventures in R&D, on the condition that firms involved sell the products of such research openly and competitively. This method of allowing collusions at the "research stage" on the condition that it is followed by competition at the production stage is often called "semi-collusion" in economic literature.

Article 86 prohibits any abuse by one or more undertakings or industries in a dominant position within the Common Market or in a substantial part of it in so far as it may affect trade between Member States. Examples of abusive conduct comprise unfair prices, production restrictions, discrimination, and predatory pricing by a dominant producer.

For discussing the collusion problem of sharing the costs of investment in knowledge capital and R&D three basic concepts are useful. One is the concept of a noncooperative Nash equilibrium (NE), which is defined under the assumption that each firm behaves as "competitively" as possible in the sense that it maximizes its own profit individually, yet without ignoring its competitors' actions. Thus if q_i is the strategy of player i out of n players, then the vector $q = (q_1, \ldots, q_n)$ is a joint strategy for all players where $q = (q_1^*, \ldots, q_n^*)$ is a noncooperative NE joint strategy vector, such that each player i strategy q_i^* is a "best reply" to others' strategies, given the optimal strategies of all its competitors. This best reply is formed by maximizing a player's profit function with respect to its own strategy, given the optimal strategies of its competitors, which are also their best replies. A Nash equilibrium is thus a collection of *simultaneous best replies*. Thus it implies that no player has an interest to deviate from it unilaterally. The concept of noncooperative Nash equilibrium is interesting because it captures the strategic nature of oligopolistic (small n) competition. With differentiated goods these equilibria allow for different strategies as prices (one for each firm). Also a set of noncooperative Nash equilibrium strategies could imply collusive profits, when the game is not a one-priced game but repeated over time. In such a "repeated game" the players compare the discounted value of future profits they could make by deviating from a tacitly collusive outcome with that if they do not deviate. When they find it is not profitable to deviate, then the collusive outcome holds and it is itself a noncooperative Nash equilibrium. This is sometimes called "tacit collusion."

A second useful concept is that of a *subgame perfect equilibrium*. Suppose a game is played in several stages. For example, firms may invest in R&D in a first stage and then sell their products in a second stage. Thus the competition law in the EU region explicitly allows collusion in the first stage but not in the second. An equilibrium is called subgame perfect when the equilibria of the subgames considered separately coincide with the equilibrium of the entire game. A subgame is the game starting at a particular stage. When the stages correspond to successive time periods, subgame perfectness implies that strategies selected in a particular period turn out to be still the best replies when considered in the next period. Thus there is time consistency.

The third useful concept is one of "normal" or "active" competition used by the EC and the Court of Justice and analyzed in some detail by Phlips (1995) in game-theory perspective. Five important propositions are discussed by Philips in this context as follows:

1. Normal or active competition implies the freedom for each individual firm to change its prices independently. Thus price agreements serve to limit this freedom.
2. Price competition between oligopolists typically takes the form of rebates, for particular transactions on the list prices.
3. Normal competition is not compatible with simultaneous moves of transactions, of prices, or of rebates or list prices.
4. Perfect information among competitors is not only a necessary condition for collusion but also a sufficient condition, because oligopolists want to collude. Thus without market transparency about prices or quantities, colluders cannot enforce a price agreement. But with market transparency they will maximize their profits overtly or tacitly or follow the price leader's moves without delay.
5. Multilateral information transmission about current or future prices among oligopolists is direct evidence of collusion.

How should active or normal competition be enforced by a court of law? We agree with Phlips (1995) that it should be equated to a competitive Nash equilibrium concept. Active competition is a major way to get an industry out of a collusive equilibrium and move it into a Nash equilibrium. As Phlips has emphasized very strongly that to reach a competitive Nash equilibrium of a single-shot game is the best that a state antitrust policy can hope for in oligopolistic markets, which is a far-reaching proposition, given that most real-life markets are oligopolistic.

We consider now the problem of shared investment and group efficiency. We assume a Cournot-type market model where each firm or player j $(j = 1, 2, \ldots, n)$ maximizes profit

$$\pi_j = p y_j - c(y_j; k_j),$$

where the market clearing price $p = a - b\sum_{j=1}^{n} y_j$ is assumed to be given and $c(y_j; k_j)$ is the long-run cost function with k_j as capacity (i.e., R&D investment) meaning in terms of output. In the short run k_j is fixed and assuming linearity the cost function may be written as $c_j = \gamma_{0j} + \gamma_{1j} y_j$. We consider now the capacity variable k_j and assume that the marginal cost of γ_{1j} depends on the level of k_j, i.e., it declines as capacity investment increases. In the long run, the capacity varies at a cost $F(k_j)$. Assume $\gamma_{1j} = \gamma_{1j}(k_j) = v_j/k_j$ and $F(k_j) = g_j \ln k_j$ with $g_j > 0$. The short- (π_j^{S}) and long-run profits (π_j^{L}) may then be written as

$$\pi_j^S = py_j - \gamma_{0j} - \gamma_{1j} y_j$$
$$\pi_j^L = py_j - \gamma_{0j} - y_j(v_j/k_j) - g_j \ln k_j.$$

On maximizing these profits one could determine the optimal outputs and capacity in the short and long run. The optimal capacity output is easily computed as

$$k_j^* = (v_j/k_j)u_j^*.$$

Now consider the situation when the n firms agree to pool their fixed capital investments in a common network $K = \sum k_j$, where the new fixed cost function is $F(K)$ with each firm's share as $\theta_j F(K)$ where $\sum \theta_j = 1$, $\theta_j \geq 0$. The joint cost function $F(K)$ is assumed to have the feature of subadditivity and economies of scale in the sense

$$\sum_{j=1}^{n} F(k_j) \geq F\left(\sum_{j=1}^{n} k_j\right).$$

This assumption is appropriate for many high-tech industries today like communications, pharmaceuticals, and computer softwares. If $\theta_j = 1/n$, then we have equal sharing and then long-run profit maximization for firm j yields the optimal capacity output as

$$k_j^{**} = (n^2 v_j/g_j)y_j^*.$$

On comparing with the optimal capacity k_j^* where the firms do not pool their fixed investment we obtain $k_j^{**} > k_j^*$ for all $n > 1$. Since the marginal cost of capacity investment is decreasing we get the result that the pooling resources and sharing total cost equally would increase the optimal capacity and output.

Sharing information in R&D research is increasingly being adopted by the high-tech firms today to keep updated about the latest developments in modern technology. Pooling of such information in order to exploit scale economies and internalize the external economies has been actively supported by the competition policy pursued by the governments in the EC.

Now consider the game-theory perspective of the competitive Cournot–Nash equilibrium of a single-shot game which is viewed as the best policy that the state antitrust policy can hope for in oligopolistic markets. Here we consider in some detail Selten's formulation analyzed by Philips. Assume a symmetric Cournot model with the same cost function for each firm as

$$C_j = \gamma_0 + \gamma y_j,$$

where the price is given by the invest market demand as before

$$p = a - bY, Y = \sum y_j.$$

To simplify the algebra we assume normalization $b = 1$, $a = 1 + \gamma$. The profit margin is then

$$\pi = a - bY - \gamma = 1 - Y.$$

The gross profit of firm j, ignoring the fixed costs γ_0, is then

$$\pi_j = y_j \pi.$$

It is assumed that fixed costs are all sunk. Given these cost and demand functions the n firms play a game comprising three successive stages as follows:

1. "A participation decision stage" in which the firms decide whether they want to participate in the bargaining to reach a cartel agreement
2. "A cartel bargaining stage" during which quota proposals are made which may either lead to a cartel agreement or not
3. "A supply decision stage" in which each firm selects a supply quantity y_j.

These stages are modeled as follows.

The "participation decision" is the selection of a zero–one variable z_j, i.e., $z_j = 0$ means no participation and $z_j = 1$ denote participation. Each firm has to decide without knowing the decisions of other firms. Let Z denote the set of all firms who select $z_j = 1$. At the end of this stage, Z is made known to all players, who are thus perfectly informed about who is going to participate in the next bargaining stage.

In the "cartel bargaining stage" each participant $j \in Z$ proposes a quota system $Q_j(q_{ji})$ $i \in C$ where Q_j is a vector of quotas or quantities to be produced by j belonging to a particular coalition C, which is a subset of Z. All such proposals are made simultaneously without knowing the proposals of other participants. A binding cartel agreement is reached when all members of a coalition propose the same vector of quotas. In the "supply decision stage" each player j selects independently a supply quantity y_j such that $0 \leq y_j \leq q_j$ where q_j is the agreed quota when j joined a cartel.

Following Philips we assume the subgame perfect equilibrium as the solution concept, so that the game has to be solved backward. First, solve the supply decision subgame for given $q = (q_1, \ldots, q_n)$ and Z. Then solve the cartel bargaining subgame, for given Z, using the equilibrium supply decisions for stage 3. Finally, determine Z using all previous subgame equilibrium results. This procedure guarantees that the solution to a subgame is also the solution to the entire game.

We start first the supply decision subgame in stage 3. If no agreement was reached, then the individual supply decisions are simply the Cournot–Nash equilibrium quantities, i.e.,

$$y_j = 1/(n+1) \ \forall j = 1, 2, \ldots, n$$

with profits $\pi_j = 1/(n+1)^2$.

If there was an agreement, then the quotas may be binding for some firms but not for others: one has to find out for each firm what the best reply quantity is, given the quota allocation. The reaction function of firm j given its best reply quantity is, given its competitors, $Y_j = \sum_{i=1}^{n} y_i \ (i \neq j)$. With a given quota vector y, this reaction function is

$$R_j(Y_j) = \max\left[0, \left(\frac{1-Y_j}{2}, q_j\right)\right].$$

This specifies the best reply of player j to the production of his competitors.

Now consider stage 2: the cartel bargaining subgame. Here the solution depends on the number of nonparticipators $N - Z$. In this second stage, k is given since it was determined by the first stage. The problem is to find out which quotas will be agreed on by the participators.

Consider the nonparticipants first. We know that their outputs are not restricted by quotas. Their best reply is $y_j = R_j(Y) = 1 - Y$. This total supply is

$$Y_{N-Z} = j(1 - Y_z - y_N - z) = \frac{k}{k+1}(1 - Y_z),$$

where $Y_z = \sum_{j \in Z} y_j$ and $Y_{N-Z} = \sum_{j \in N-Z} y_j$.

Now consider the participators. They supply a quantity $1 - Y$, if that is smaller than their quota or else their quota. So $y_j \leq 1 - Y$ for $j \in Z$ and their total supply y_Z is $Y_Z \leq (n - l)(1 - Y)$. Finally, the joint profit of participators $\pi_Z = \sum_{j \in Z} \pi_j$ can be written as

$$\pi_Z = (k+1)^{-1} Y_Z (1 - Y_z).$$

Clearly this joint profit reaches its maximum at $Y_Z = \frac{1}{2}$.

It can be shown that the number of nonparticipators k plays a crucial role in the success of a cartel agreement. For instance if $k > (n-1)/2$ a quota agreement has no effect in the sense that the participation will behave like nonparticipators.

Now consider stage 1: the participation decision subgame. It remains to find out how many players participate in a cartel agreement. Selten has shown that if there are up to $n = 4$ firms in a market, for all of them to participate is a subgame perfect equilibrium. That is why four are said to be few. When there are any more than four firms, cartel equilibria with less than the total number of firms participating can be found by calculations. Philips has analyzed this calculation in some detail. However, if the number of nonparticipators is $k > (n-1)/2$, then every player receives the unrestricted Cournot–Nash equilibrium profit.

Thus we consider "active competition" as one way to get an industry out of a collusive equilibrium and move it into a competitive Nash equilibrium. This equilibrium defines the lower limit to which active competition should reduce industry prices or the upper limit to which active competition should push industry production. Once this limit is reached, no oligopolist has an incentive to break through it.

We now consider the problem of sharing technology and sharing of innovation costs through discriminatory pricing. We explore in this section two types of models discussed by Baumol (2002) in some detail. One is the technology consortium model, which deals with the market process of sharing new technology and innovation. The sharing process helps the innovating firms in several ways, e.g., it helps to internalize some or all of the spillover effects, reduces uncertainty of R&D investment, and avoids duplication of research cost. It also increases the scale of research, where fixed cost is very large. Two institutional arrangements help this process. One is that the anti-monopoly laws in most capitalistic countries do not allow collusion in products and services but allow collusion in R&D because it has a public good character. Secondly, the sharing allows consumers' surplus to increase since higher scale reduces unit costs of R&D.

Baumol has discussed five major reasons why firms build joint ventures for R&D projects with more cooperation than rivalry:

1. Firms gain a competitive advantage when firms pool their resources.
2. Each firm in a technology consortium has a strong incentive through present cost reduction and expectation of future benefits with its agreements, giving full access of all information to the partners.
3. The consortium also helps in stimulating future innovations.
4. It also helps to increase consumers' surplus for the whole economy through scale effects and price reductions.
5. The consortium eliminates all potential losses from infighting and competition among firms when they do not form the joint venture. There exist both complements and substitutes in innovation. The consortium can help eliminate substitutes and augment the complements in the innovation process.

The consortium model has two parts. One emphasizes the point that the firms which exchange R&D information with other members of the consortium are more profitable than those which do not join the joint venture. The second part shows that if each firm in the consortium behaves like a Cournot oligopolist and there is complementarity among the research output of the technology sharing firms, then a rise in the number of consortium members will increase each member's outlay on innovation, as well as the output of the total product, and shift its total cost function downward. For the second part of the consortium model we consider a simple derivation, where each symmetric Cournot firm j $(j = 1, 2, \ldots, n)$ maximizes profit

$$\pi_j = py_j - C(y_j, k_j)$$
$$p = a - b \sum_{j=1}^{n} y_j,$$

where p is the market clearing price and $C(\cdot)$ is the cost function depending on output y_j and capacity k is measured in terms of output. In the short run k_j is fixed and assuming linearity the cost function may be written as

$$C_j = h_{0j} + h_{1j} y_j \ \forall y_j \leq k_j.$$

In the long run the capacity output variable also varies at a cost $F(k_j)$. Assume that the marginal cost h_j depends on the level of k_j. In many high-tech industries like computers and electronics this marginal cost $h_{1j}(k_j)$ declines as capacity is increased. Thus the capacity expansion gives rise not only to economies of scale but also to lower variable cost. This capacity variable may be a proxy for R&D knowledge and innovation capital. It builds dynamic core competence of high-tech firms. We may represent the marginal cost function as $h_{1j}(k_j) = \frac{v_j}{k_j}$ and $F(k_j) = g_j \ln k_j$ with $g_j > 0$. The long-run profits then become

$$\pi_j^{\mathrm{L}} = \left(a - b\sum_{j=1}^{n} y_j\right) y_j - h_{0j} - y_j \frac{v_j}{k_j} - g_j \ln k_j,$$

whereas the short-run profit π_j^{S} is however given by

$$\pi_j^{\mathrm{S}} = \left(a - b\sum_{j=1}^{n} y_j\right) y_j - h_{0j} - y_j.$$

With the short-run optimal output

$$y_j^*(S) = \frac{1}{2b}\left(a - h_{1j} - b\sum_{j=1}^{n-1} y_j\right).$$

The long-run optimal capacity output k_j^* is obtained by setting the derivative of π_j^{L} with respect to k_j equal to zero, i.e.,

$$k_j^*(L) = \frac{g_j}{v_j} y_j^*(L).$$

Now consider the situation when the n firms agree to pool their fixed capital investments or R&D capital in a technology consortium $K = \sum k_j$. With the new cost function $F(K)$ where each firm's share is $\theta_j F(K)$ with $\theta_j \geq 0$ and $\sum \theta_j = 1$, the joint cost $F(K)$ has the feature of subadditivity and economies of scale in the sense

$$\sum_{j=1}^{n} F(k_j) \geq F\left(\sum_{j=1}^{n} k_j\right).$$

These features are appropriate for many high-tech industries today. If $\theta = \frac{1}{n}$, then we have equal sharing and the long-run profit function then becomes

$$\pi_j^{LR} = \pi_j^{\mathrm{L}} - \frac{1}{n}(g_j \ln k_j).$$

Its maximization yields the optimal capacity as

$$k_j^{**} = \frac{n^2 v_j}{g_j} y_j^*.$$

If however the firms do not pool their R&D capacity in a consortium, then we get

$$k_j^* = \frac{v_j}{g_j} y_j^*$$

Clearly for $n > 1$ we obtain $k_j^{**} > k_j^*$ where y_j^* equals $y_j^*(S)$ derived above.

Despite the loss of profit resulting from exclusion from a technology consortium, Baumol shows that it does not follow that incentives for cheating are absent. Such incentives do exist for technology agreements and sharing. However the information-exchange cheating is apt to be discovered eventually and the firm that does cheat is likely to be deprived of the benefits of membership. There also exist other formal arrangements for discouraging cheating. Thus technology consortia are relatively immune from destabilizing cheating. Another important model developed by Baumol discusses the issue of recoupment of innovation costs most of which are sunk cost. The need to recover continuing and repeated sunk costs leads to discriminatory pricing in the oligopolistic innovation industries. Baumol advances three propositions in this framework:

1. Where discriminating pricing is possible (e.g., when the demand curves of different customer groups have different price elasticities) there will always be a set of discriminatory prices for a given product that yields higher profits than any uniform price.
2. Zero-entry barriers will preclude positive economic profits, but they will not prevent incumbent firms from covering all of their costs, including common costs, fixed costs, and continuing sunk costs.
3. In an oligopoly market that is perfectly contestable, Ramsey prices are sustainable against entry.

Baumol showed that discriminatory pricing itself always seems to attract niche entrants who skimp on the sunk costs that would enable them to compete with full effectiveness in the long run. But overall this sort of pricing is essential to cover the continuing sunk costs. He formalized this as follows:

Proposition: If the marginal cost curves of all the firms are identical and U-shaped, and where industry output is not an integer multiple of y_m, the optimal output which minimizes AC (average cost), and economic efficiency requires the output of every firm to deviate equally from its AC-minimizing level. A corollary is that for efficiency, the deviation of the output of the firm from the AC-minimizing output to be a decreasing function of the number of firms in the industry.

Two exceptions are to be noted. One is that firms are not identical in their AC curves due to difference in sizes. There will be superior firms where all efficiency rents would go to the inputs responsible for a firm's superior performance. Secondly, under discriminating pricing of a limit pricing model, the dominant firms may earn extra rents in a framework where the leader follower network prevails.

1.5 Economic Implications

The innovation models we have discussed analyze firm growth through efficiency and their dynamic impact on industry growth. On its part industry growth generates overall economic growth. In recent times countries of Southeast Asia, e.g., South Korea, China, Taiwan, and Singapore, have achieved remarkably high growth rates over the last two decades or more, and to a large part this growth rate has been achieved through successful adoption of incremental innovations and modern technology borrowed from the advanced industrial countries. These newly industrializing countries (NICs) in Asia have stressed on expanding their knowledge capital and captured the dynamic scale economies and allocative efficiency and competed very successful in world markets.

Baumol emphasized three major reasons why firms build joint ventures in research and innovation network:

1. The cumulative character of many innovations, e.g., they add to new technology and the spillover effects allow diffusion of technology through R&D network.
2. The well-known public good property of new information technology which contribute to the output not only of the firm that made the breakthrough but also of other firms.
3. The new innovation with its network effects has steady-state growth effects, not simply the level effect. This has been called the accelerator feature of most modern innovations.

The NICs of Southeast Asia effectively utilized this growth-enhancing effects of incremental innovation and achieved very high growth rates of their output and income. Sengupta (2010) has recently discussed in some detail the impact of new technology and innovation in these high-performing economies, which are playing a most dynamic role in the world market today.

Three important creative processes played a catalytic role in these high growth NICs in Asia.

The first is the role of private enterprise sector, where the state provided active support in various ways. Thus China's reform of the national innovation system since 1990 emphasized the role of the enterprise sector. Thus in 2000 60 % of China's R&D spending was performed by the enterprise sector. Most of the enterprise-funded R&D was performed outside the state-owned enterprise system. A good measure of R&D intensity is the ratio of R&D expenditure to GDP. By this measure China's R&D intensity rose from 0.74 in 1991 to 1.23 in 2003 and 1.89 in

2008. For South Korea it rose from 1.92 in 1991 to 2.96 in 2003. For Taiwan it rose from 0.82 in 1991 to 2.16 in 2003 and has exceeded 2.75 in 2008.

Technology diffusion through learning by doing has been the second most important factor in the innovation dynamics of NICs in Asia. Taiwan's contemporary knowledge-based economy has revealed a more remarkable growth of the ICT (information and communication technology) than China and other NICs of Asia. From 1995 to 1999, Taiwan's ICT industry ranked third in the world after USA and Japan. The World Economic Forum (2006) has computed a GCI based on institutions and the adoption of best practice technology. Its report for 2003 shows Taiwan's rank to be fifth, while Japan and South Korea had 11 and 18, respectively. Here the state took significant initiatives encouraging the high-technology firms to augment their R&D investments and establishing special zones such as Hsinchu Research Park, where agglomeration and skill complementarities were utilized. One measure of inventiveness in Taiwan is its record of US patent awards. In 2003 for example Taiwan had the average annual number of US patents per million people as 241 with rank 3 with USA and Japan holding first two ranks. A national innovative capacity index constructed by Porter and Stern (2004) showed Taiwan's position at 32.84 while USA and Japan were at 36.60 and 34.62. Taiwan exceeded South Korea (31.13), China (25.86), Malaysia (26.85), and India (25.52). We have to note also that Taiwan utilized the linkage with small and medium industries to foster technology diffusion most rapidly.

Finally, the innovation efficiency in the NICs had achieved a steady rate of increase over the years through its four components: learning efficiency, technical and allocative efficiency, and scale economies through joint ventures and network effects. Recently Lopez-Claros and Mata (2010) constructed a composite measure called the innovation capacity index (ICI) based on a weighted average of five pillars as they are called:

1. Institutional environment which includes among others public sector management, corruption perception index, and the state of the macroeconomy.
2. Human capital, training, and social inclusion, which include among others adult literacy, secondary and tertiary gross enrollment ratio, and health worker density.
3. Regulatory and legal framework including investment climate and administration of tax policies.
4. Research and development which includes R&D worker density and patents and trademarks.
5. Adoption and use of information and communication technologies, which include among others the use of mobile phones and Internet, government's use of ICT, and electrification rate.

This ICI is more comprehensive than other similar indexes constructed by OECD. This ICI in its 2009 version covers 131 countries and identifies over 60 countries that are seen to have a bearing on a country's ability to create an environment that will encourage innovation. Some rankings for this index over 2009–2010 are as follows:

Country	ICI rank	ICI score
Sweden	1	82.2
USA	2	77.8
Singapore	6	76.5
Taiwan	13	72.9
Japan	15	72.1
Hong Kong	16	71.3
South Korea	19	70
Malaysia	34	57.3
China	65	49.5
India	85	45.6
Brazil	87	45.2

Clearly the NICs in Asia, e.g., Singapore, China, Taiwan, South Korea, and Hong Kong, fare very well in the ICI. Although ICI is a very rough measure it has strong emphasis on R&D and macroeconomic policies pursued by different countries.

To conclude this chapter we note that innovations in different forms are going to play a most active role in industry growth and world trade for the future. This may disrupt some of the equilibrium assumptions of the competitive general equilibrium models of the neoclassical school. The role of information sharing and technology consortium is going to occupy a most critical phase in the world economic framework. New theory must be applied to analyze these latest trends where innovations occupy the center stage.

Chapter 2
Innovation Models

Models of innovation emphasize for the most part the endogenous character of industry growth. How does a firm grow? How does it affect industry growth? What role does the market play in this framework? These issues are central to the formulation of innovation models in recent times. Innovation models vary by types of innovations and the way they affect industry growth. However most innovations have some common characteristics as follows:

1. It involves new and productive ways of industry growth through production, distribution, communication, or organization.
2. R&D investment comprises the core component of most innovations, and it may involve both theoretical and applied research. However theoretical research does not directly yield industrial results till it is applied through knowledge diffusion across firms and industries commercially.
3. Technical change and the diffusion of human capital play a central dynamic role in most industrial innovations, although creating a new firm or a new organization may be equally important for starting an innovation.
4. All endogenous innovations are motivated by the market incentives of profit and economic efficiency under dynamic competition.

In order to analyze these common characteristics of innovation we have to discuss the innovation diversity. This diversity involves various forms of innovations, which are all dynamic in character in the sense that they have economic impact and evolution over time.

2.1 Innovation Diversity

Innovations may take many forms depending on the gestation period of developing it and on its evolutionary impact over time. Four broad types are usually distinguished:

J. Sengupta, *Theory of Innovation: A New Paradigm of Growth*,
DOI 10.1007/978-3-319-02183-6_2,

1. Technology-based innovations
2. Knowledge diffusion and human capital based innovations
3. Introduction of "new combinations" which creates a fundamental impulse in capitalist development to generate new consumers' goods, new methods of production or transportation, new markets, or new forms of industrial organization
4. Finally, innovation as evolutionary learning through adaptive efficiency in dynamic competition and/or Cournot–Nash market structure.

It is useful to discuss these four varieties of innovations, since they involve some of the basic features of modern industry growth in several fields such as information and communication industries, pharmaceuticals, and bioengineering fields.

Technical change in the form of new techniques of production or new processes or software development is mostly used to characterize technical inventions which alter the shape of the production and cost frontiers of firms and industries. In Solow's model of neoclassical growth this comprises the major form of technological progress. The technology process comprises several stages. The first is technology creation and diffusion. Research and its interactions through diffusion of both theoretical and applied knowledge are important here. An empirical estimate by Cohen and Levinthal (1989) over 1,302 sample units in the US manufacturing sector shows that the effects of innovations on R&D intensity of the basic and applied sciences are significantly different. This means that the role of diffusion and learning differs significantly across industry fields. Increasing technological opportunity through the basic sciences evokes more R&D spending than does increasing technological opportunity through applied sciences. The second important aspect of the new technology process is its impact on new types of dynamic efficiency such as innovation efficiency and access efficiency. Innovation efficiency occurs through the competitive race in the knowledge arena. The drive for imitation and the first mover advantages provide the incentives for capturing the industry spillover effects. Access efficiency occurs through globalization of markets through mergers and technology consortium. Finally, new technology changes the market structure dramatically, especially in the high-tech fields, e.g., through software development and miniaturization of new technical gadgets.

Although new technology is more visible in its physical impact, the human capital based innovation has more long-run impact on the overall economic growth of nations. In recent times competition has been most intense in modern high-tech industries such as semiconductors, microelectronics and personal computers. The empirical study by Jorgenson et al. (2000) noted two significant impacts of the growth of computer technology on the overall US economy. First, as the computer quality improves, more computer power is being produced from the same inputs, i.e., learning by doing through human capital and cumulative experience on the job increases the skill inputs. Secondly, the computer-using industries are now using skilled labor working with better and more efficient computer equipment and the related communications equipment like the iPad, thus increasing labor productivity in the high-tech and other manufacturing and service industries. For example, the average industry productivity growth (i.e., total factor productivity growth) has achieved a rate of 2 % per year over the period 1958–1996 for electronic equipment, which includes semiconduction and communications equipment. This trend has continued over the recent period 2000–2010 although at a slightly lower rate. High productivity growth led to falling unit costs and prices. For instance, average PC prices declined by 18 % per year from 1960 to 1995 and by 27.6 % over 1995–1998. Recent estimates suggest a rate of decline of 15 % over 2000–2010. Learning by doing through the use of innovations in human capital has greatly contributed to this high growth phenomena. We consider now three types of learning phenomena involving knowledge diffusion. One is the cumulative research experience embodied in cumulative output, where the latter is very often used as a proxy measure of technological progress. The second measure is cumulative experience embodied in strategic inputs such as capital goods or R&D inputs. Finally, the experience in "knowledge capital" available to a firm due to a spillover from other firms may be embodied in its cost function through cumulative research inputs.

In order to characterize efficiency through human capital utilization we use nonparametric efficiency models through a set of linear programming models, also known as DEA (data envelopment analysis) models. These models characterize Pareto efficiency and screen the efficient firms over the inefficient ones. The unifying theme of these models is a convex hull method of characterizing the production frontier without using any market prices and also the cost frontier which uses market prices to determine the optimal level of inputs. Consider the problem of testing the relative efficiency of a reference firm h in a cluster of N firms, where each firm produces s outputs (y_{rj}) with two types of inputs: m physical inputs (x_{ij}) and n human capital inputs as knowledge capital (z_{wj})

$$\min u + v$$

subject to (s.t.)

$$\sum_{j=1}^{N} x_j \lambda_j \leq u x_h; \sum_{j=1}^{N} Z_j \lambda_j \leq v Z_h \tag{2.1}$$

$$\sum Y_j \lambda_j \geq Y_h; \sum \lambda_j \geq 1, \lambda_j \geq 0$$

$$j = 1, 2, \ldots, N.$$

Here x_j, Z_j, and Y_j are the observed input and output vector for each firm j. Let $\lambda^* = (\lambda_j^*), u^*, v^*$ be the optimal solutions of the LP model (2.1) with all slack variables zero. Then the reference unit or firm h is technically efficient (without using any market prices) if $u^* = 1.0 = v^*$. If however u^* and v^* are positive but less than unity, then it is not technically efficient at the 100 % level, since it uses excess inputs $(1 - u^*)x_{ih}$ and $(1 - v^*)z_{wh}$. Overall efficiency (OE_j) of a firm combines both technical (TE_j) or production efficiency and the allocative (AE_j) or price efficiency as follows: $\mathrm{OE}_j = \mathrm{TE}_j * \mathrm{AE}_j$. For overall efficiency one may solve the cost minimizing model:

$$
\begin{aligned}
&\min \mathrm{TC} = c'x + q'z \\
&(\text{s.t.})\ X\lambda \le x;\ Z\lambda \le z \\
&Y\lambda \le Y_h;\ \lambda' e = 1, \lambda \ge 0.
\end{aligned}
\tag{2.2}
$$

Here e is a column vector with N elements, each of which is unity, prime denotes the transpose, c and q are unit cost vectors of the two types of inputs x and z which are now the decision variables, and $X = (X_j)$, $Z = (Z_j)$, and $Y = (Y_j)$ are appropriate matrices of inputs and outputs. Overall efficiency (OE_h) computed from (2.2) is $\mathrm{TC}_h^*/\mathrm{TC}_h$; the allocative efficiency is $\mathrm{AE}_h = \mathrm{TC}_h^*/(u^* + v^*)\mathrm{TC}_h$ where TC_h and TC_h^* are the observed and optimal costs for firm h. Clearly the inefficient firms have minimal total costs with maximum overall efficiency. These firms would tend to lead in industry growth and attain a more dominant position.

Now consider the special characteristics of the research inputs z associated with human capital. Since these inputs tend to affect unit costs nonlinearly we can rewrite the objective function of model (2.2) as

$$
\min \mathrm{TC} = \sum_i \left[\left\{ c_i - f_i \left(\sum_w q_w z_w \right) \right\} x_i + \frac{1}{2} d_i x_i^2 \right] + \left(\frac{1}{2} \right) \sum_{w=1}^{n} g_w z_w^2 . \tag{2.3}
$$

Subject to the constraints of the model (2.2). Here f_i is the unit cost reduction with $f_i < c_i$ and the component cost functions are assumed to be strictly convex. If the firm is efficient with positive input levels and zero slack variables, then we must have $\partial L/\partial z_w = 0 = \partial L/\partial x_i$ where L is the Lagrangian function. The Kuhn–Tucker necessary conditions for optimality are then

$$
\begin{aligned}
f_i q_w x_i + \gamma_w &\le g_w z_w;\ z_w \ge 0 \\
f_i \left(\sum q_w z_w \right) + \gamma_i &\le c_i + d_i x_i;\ x_i \ge 0.
\end{aligned}
$$

Here γ_i and γ_w are the Lagrange multipliers for the first two constraints. Here the complementarity of the two inputs x and z is explicitly shown and the learning parameter f_i captures the productivity impact of human capital.

A more general version of the model arises when we incorporate the time profile of output generated by the cumulative experience through human capital investment. Let $z(t) = (z_w(t))$ be the vector of gross investments and $k(t) = \int_0^t z(s)ds$ be the cumulative value where

$$
\dot{k}_w(t) = z_w(t) - \delta_q k_w(t) \tag{2.4}
$$

with dot denoting the time derivative and δ_w is the fixed rate of depreciation. The long-run cost minimization model now becomes

$$\min J = \int_0^\infty e^{-rt}\left[c'(t)x(t) + C(z(t))\right] \mathrm{d}t \quad (2.5)$$

s.t. (2.4) and the constraints of (2.2).

Here $C(z(t))$ is a scalar adjustment cost which is generally assumed to be nonlinear in the current theory of investment. On applying Pontryagin's maximum principle, the optimal long-run cost frontier of the efficient firm can be determined, which can explicitly show the interdependence of the two inputs: the physical and human capital based research inputs.

Introduction of new combinations of inputs and creation of new markets have become a major source of high-tech development in the field of communications. Miniaturization and multifunction-based features have currently dominated the iPhones, iPad markets and competition is very intense here, where different types of technologies are converging. Drawing on a data set of more than 2,000 observations on "significant innovations and innovating firms" in the UK over the period 1945–1980, Pavitt (1984) identified five major categories of innovations as follows: (1) supplied dominated, (2) production intensive, (3) science based, (4) information intensive, and (5) technological trajectory based. Lancasterian and post-Lancasterian characteristic-based approach to the definition of product represent a powerful tool to operationalize a set of innovation modes. Innovation is viewed here not as an end result but as an ongoing process. Rather than identifying "types" of innovations, this framework allows us to identify different "models" of innovation and their dynamics. Some of these dynamics are as follows:

(a) *Radical innovation*: It is defined by the creation of a new set of vectors of competencies, technical and service characteristics.
(b) *Improvement innovation*: It occurs when the set of vectors of characteristics remains unchanged, but the quality value of their single elements increases through technical characteristics or improvements in certain competence vectors.
(c) *Incremental innovation*: This occurs when a new characteristic of the product or process is added or modified. It is not the traditional notion of a sort of residual.
(d) *Ad hoc innovation*: From the supplier viewpoint this innovation means contributing to the whole set of competencies, making significant change in the vector of competencies, and improving the immaterial knowledge elements of the technical characteristics vector.
(e) *Recombination innovation*: This might involve creation of a new product or new process as a combination of characteristics of one or more products, or through fragmentation of the characteristics of a preexisting product or process.
(f) *Formalization innovation*: This occurs when one or more characteristics of new products or processes are formatted or standardized. This occurs most frequently in the service-oriented industries and communication industries.

Evolutionary learning and spillover knowledge across firms, industries, and nations emphasize the most creative role of innovation dynamics. This is supremely important in today's world of high technology. Modern endogenous growth theory assumes that an economy automatically benefits from its investments in new knowledge, because knowledge is a public good that can be used by an entire economy. The USA is thought to commercialize new knowledge better than Europe. The successful NICs in Asia have performed much better than Europe and Latin America. Investment in new knowledge is only a necessary condition for endogenous growth: this new knowledge must be exploited and put to commercial use so that it can translate into stronger competitiveness and cumulative economic growth. The contribution of Acs et al. (2009) has extended the microeconomic foundations of the macro models of endogenous growth through the knowledge spillover theory of entrepreneurship, which holds that knowledge creation can lead to knowledge spillovers, creating technological opportunities. These opportunities can be exploited by new entrepreneurs and businesses. New product innovations may be generated from both incumbent firms and start-ups. The incumbent firms may produce incremental innovations from the flow of knowledge, whereas start-ups may exploit knowledge spillovers to produce major or radical innovations. One major impact of spillovers and external economies of knowledge creation is that it generates significant scale economies which spread across national boundaries. By the comparative advantage principle of international trade theory, the size becomes an important factor. Through mergers and acquisitions businesses exploit the advantage of market expansion throughout the world.

Thus innovations tend to provide many channels of potential market power, which challenges the basic premises of competitive equilibrium. In the present day world of innovations and their spillover effects various forms of noncompetitive market structures have evolved in recent times and correspondingly many dynamic models of innovations have been formulated. Scale economies and learning by doing have comprised the basic elements of such innovation models. Economies of scale are usually measured in two ways: either through a production function showing increasing returns to scale or a total cost function showing declining long-run average cost. Four major sources of economies of scale have been distinguished in economic literature: (1) indivisibility of specific inputs like the plant or knowledge capital sometimes measured by the size of the plant or capital stock, (2) learning by doing through cumulative experience embodied in labor or human capital, (3) the industry stock of R&D and knowledge capital, and (4) economies of scope due to integration of managerial and technical functions. External economies (EO) are important for two reasons. One is that it captures the effects of technological diffusion across firms and related industries. This spread effect may arise through both output of EO and its linkage effect through complementarities. Knowledge, e.g., software technology, helps other firms grow. Consider a linear cost function for firm j as $c_j = ay_j$, where y_j is output. The industry cost function in the symmetric case is $C = aY$, where $Y = \sum_{j=1}^{n} y_j$ and $C = \sum_{j=1}^{n} c_j$. The industry effect is then captured by relating the constant a in the firm cost function to the industry output as

$$a = a_0 Y^{r-1}, a_0, r > 0.$$

In a competitive framework the individual firm equilibrium is given by $p = MC = a$, which is subjective to the firm. The objective condition of equilibrium however allows for external economies and is given by $p = arY^{r-1}$. If r is positive but less than one, then the equilibrium price will fall as total industry output increases. Also the marginal cost of each firm would fall as total industry output rises. The price fall would generate market expansion and global trade would be augmented through diffusion of knowledge and external economies. The major factor in technology and knowledge diffusion is the interindustry spillover in the high-tech industries today. Bernstein and Nadiri (1988) estimated the effects of these spillovers on production processes by specifying a variable cost function for each industry as a truncated translog function with no own of squared second-order terms:

$$\ln(c_v/w_m) = f(\ln w_l, \ln w_p, \ln y, \ln K_t^i, \ln K_t^j, \text{other terms}).$$

Here c_v is variable cost, w_m = price of materials, w_l = wage rate, w_p = rental rate on capital, y = output, K_t^i = industry's R&D capital, and K_t^j is other industry's R&D capital. The empirical estimates for Us industries such as electrical products, scientific instruments, chemical products, and transportation equipment showed two general findings. One is that the variable cost for each high-tech industry was significantly reduced by R&D capital spillovers. Secondly, the technology industries had higher rates of cost reduction, e.g., computers and telecommunication industries.

Due to these spillover effects and external economies various forms of noncompetitive market structures have developed in recent times in these high-tech industries. Following Schumpeter's innovation approach D'Aveni (1994) has characterized this noncompetition framework as hypercompetition. Whereas the competition paradigm emphasizes pricing as the basic strategy with a fixed technology, hypercompetition stresses the dynamics of innovation in both technology and market structure. This is very similar to Schumpeterian innovation theory, where innovations shift the production and distribution frontier and the opportunity to make quasi-monopoly profits through the innovation process provides the basic endogenous motivation for increased investment for human capital. The shift from perfect competition to noncompetition market structure brought about by innovations allows market rivalry and increased market entry for the successful innovations. By innovations the firms tend to occupy a dominance in the market structure. This dominance creates two types of impact on industry performance. One is the leader–follower interdependence analyzed in the Stackelberg model and the second is the entry preventing strategy adopted by the successful innovators as dominant firms.

2.2 Innovation Models of Industry Growth

Dynamics of innovation efficiency characterize the process of industry growth through a number of factors such as new technology, new source of supply, and/or a new type of organization or communication network. We discuss here four specific types of models of innovation dynamics as follows:

1. A quasi-competitive model of innovation
2. A model of technical innovation and diffusion
3. A dynamic Cournot–Nash model with spillover effects
4. A demand-induced model of innovation

The relationship between innovation and industrial evolution has always been central in Schumpeterian work on innovation. Recent trends have extended this discussion in several ways. We discuss here a few of these trends:

(a) In some models such as Sutton (1998), technology and demand-related factors set bounds on industrial structures, entailing Nash equilibrium on industry-specific entry processes. No specific attention is paid to the learning process of firms.
(b) The competitive process in some recent models of industry dynamics becomes proactive in weeding out the heterogeneity in firm distributions.
(c) More intensive on the learning processes of firms in industrial dynamics are the evolutionary models of Nelson and Winter (1982) and Dosi et al. (1995).
(d) Progress has been made at a more macrolevel by linking innovation and industry evolution to structural change and the changing structural composition of the economy. The work of Metcalfe (1998), Dosi (2001), Montobbio (2002), and others. DeBresson and Andersen (1996) and his associates have empirically estimated an innovative-interaction matrix between sections which are suppliers of innovative activity and the sectors which are users.

For discussing these models we follow separate notations in each case according to each author or authors.

A. A Quasi-Competitive Model

We consider a short version of the competitive model of innovations due to Amendola and Musso (2000) and discuss the noncompetitive features underlying it, which make it quasi-competitive.

The model shows that competition operates through innovation, where innovation is a means of reducing production and distribution costs or for capturing new markets. Innovative choices do not consist in the instantaneous development and adoption of new technologies, new products, and/or new organizational forms. They involve in general the substitution of the new for the old and this takes time. Hence the problems of coordination naturally arise. The Walrasian adjustment mechanism in competitive framework assumes various stringent equilibrating processes for demand supply adjustments through price and cost changes. The selection mechanisms provide some alternative methods of relaxing the stringent assumptions of the perfectly competitive model. Consider for example the following situations: (1) firms must pay a sunk cost to enter the market; (2) all firms do not have the same market information; (3) not all firms have the access to the same technology; and (4) buyers do not have the same information about the different sellers. In each of these cases some assumption of perfect competition is violated. Two consequences of this imperfect situation are that different firms would have different levels of cost efficiency and firms would attempt to learn about their own efficiency through competitive market signals. The competitive selection model implies that different firms earn different profit rates even in the long run. By competitive selection the firms that remain active and efficient have a level of efficiency that is higher than the average. Finally, the quasi-competitive selection model is consistent with the empirical situation that the firm size distribution is neither single valued nor indeterminate as the perfect competition model would imply.

The model due to Amendola et al. (2000) discusses the time structure of production processes and analyzes the sequential interaction of competing firms ($i = 1, 2$) in a process of restructuring of productive capacities.

At each time point t the production capacity of a firm i is represented by the intensity vector $X^i(t)$ and the level of activity constrained by available financial resources:

$$F^i(t) = m^i(t-1) + h^i(t-1) + f^i(t),$$

where $m^i(t-1)$ and $f^i(t)$ are internal and external financial resources with $h^i(t-1)$ as the money balances accumulated in the past. Since $m^i(t)$ is the monetary proceeds from the sale of final output and prices are fixed within each given period, one could write $m^i(t) = \min\left[p^i(t), d^i(t), p^i(t)s^i(t)\right]$ where $s^i(t)$ and $d^i(t)$ are the current real supply and demand for firm i and $p^i(t)$ is the price. Excess supply results in an accumulation of undesired stocks $c^i(t)$ for the firm. The current final output is then

$$q^i(t) = s^i(t)\theta c^i(t-1), 0 < \theta \leq 1$$

which can also be written as

$$q^i(t) = u^i(t) \sum_k B_k^i(t) x_k^i(t)$$

with u^i being the rate of utilization of productive capacity inherited from the past and $B_k^i(t)$ denotes the output coefficients of the production process. The aggregate market demand $D(t)$ is determined as

$$D(t) = (1 + \hat{g}) D(t-1) p^v, v \leq 0,$$

where p is average market price with an exogenously determined growth rate $\hat{g}$. The evolution path followed by each firm is actually determined by the behavior of the decision variables, namely, the rate of starts of new production processes x^i, the rate of utilization of productive capacity u^i, the price of final output $p^i(t)$, the wage rate, the ratio of external financial resources $f^i(t)$ to $m^i(t)$, and the rate of scrapping. Each firm determines its current utilization of productive capacity $u^i(t)$ so as to adjust its current supply to the expected final demand $\hat{d}^i$:

$$u^i(t) = \min\left[1, \frac{\hat{d}^i(t) - (c^i(t-1) - c^i_d(t))}{\sum_k B^i_k(t) x^i_k(t)}\right].$$

The price charged by each firm is determined in such a way as to cover the cost of production when using the up-to-date technology adopted at the desired rate of capacity utilization. This price can be adjusted in the Walrasian fashion in reaction of the market disequilibrium as

$$\bar{p}^i(t) = p^i(t)\left[1 + r^i \frac{d^i(t-1) - s^i(t-1)}{s^i(t-1)}\right].$$

The performance of each firm $0 \le r^i \le 1$ is measured here by its unit margins, i.e., as the ratio of the difference between the price and the current unit cost of output. Unit margins on average equal to zero means that firms realize normal profits. Unit margins would be necessarily negative at the beginning of any innovation process involving higher construction costs.

Two aspects of this model require further discussion. One is the impact of the innovation process on price competition. This is analyzed in this model by a series of simulation experiments. The second is the interaction between technology and the innovation process as it affects the Walrasian adjustment process. Competition is really successful when price and quantity adjustments are carried out which make it possible to obtain normal profits, i.e., when these adjustments do not result in waste of productive resources. Thus viewed competition not only coexists with increasing returns but helps the firms to capture them.

Four simulations are performed to simulate the impact on the innovation process of a price competition between the two firms. In the first simulation the two firms ($i = 1, 2$) innovate one after the other but with the same frequency. With no financial constraints and fixed nominal wages, both firms remain on the market. Both firms realize positive unit margins from innovations. Now if firm 1 (the first mover) innovates more frequently than firm 2, the latter exits from the market. Simulation 3 allows a larger asymmetry in innovation frequencies. This allows firm 1 to reestablish a definitive competitive advantage, which may be compensated by a larger asymmetry in price reaction by firm 2. A sustained price competition allows the less innovative firm to stay on the market. Then any first mover expecting to be confronted by a stiff price competition has very little real incentive to innovate. Simulation 4 shows that a very strict financial constraint makes it possible to reestablish the incentives required by a pure innovation strategy.

The technology innovation process is basically a qualitative change, i.e., a change which takes place through distortions of productive capacity which imply the appearance of problems of coordination between supply and demand. In order to explore how competition operates as an ordering force, it is not appropriate to consider a static world, where competing firms are making similar products in given and unchanged cost conditions. Instead it is more appropriate to envisage firms which undertake innovative activities as in Richardson (1997). The dynamics of this innovative world of competition is at the core of the Walrasian adjustment process. Dramatic changes in new technology have made the adjustment process multidimensional. For example, new products are increasingly being modularized and standardized and suppliers of components are increasingly involved in innovation. An important challenge comes from networks. This challenge starts from the recognition that innovation and industry evolution are highly affected by the interaction of heterogeneous actors with different knowledge, competencies, and specialization, with relationships that may range from competitive to cooperative, from formal to informal, and from market to nonmarket.

B. A Technological Innovation and Diffusion Model

This model developed by Iwai (1984) intends to show that under the Schumpeterian hypothesis on technological diffusion and innovation, an economy's state of technology will be in a perpetual disequilibrium. Franke (2000) has discussed this model in some detail and showed that although the evolutionary force of competition and pressure on the profit-maximizing firms steer the economy toward a neoclassical equilibrium in which all firms use the most efficient production technique available, the function of innovation is precisely to upset this equilibrating tendency. Through a series of simulations of the stochastic version of the general model he showed that the assumption of a stochastic arrival of innovations typically yields long waves of oscillations in the growth rates of average productivity. These long waves can be viewed as originating with the frequency distribution of techniques on a wave train.

The diffusion model developed by Iwai (1984) is based on three exogenous parameters: λ is the rate of technical progress, i.e., the average growth rate of productivity of newly invented techniques; the innovation parameter ν is a measure of the effectiveness with which firms introduce these inventions; and γ represents the speed of technical diffusion. There are systematic cost gaps between the existing production methods and the best practice technique (BPT) currently in use. Assuming production techniques with fixed unit costs c, the cost gaps denoted by z are specified as the logarithmic differences to the BPT unit cost c, i.e., $z = \ln c - \ln C$. While the level of the BPT unit cost decreases over time at the average rate λ, the relative frequencies of techniques with a given distance to the current BPT are assumed to be stable. The distribution of the expected values of the capacity shares in the statistical equilibrium can be described by a density function $\bar{s} = \bar{s}(z)$.

The process of technological diffusion is assumed to be governed by differential cost advantages, which occur gradually. Assuming a speed of adjustment $\gamma > 0$, the changes in capacity can normally be expressed directly as

$$s_{i,t} = [1 - h\gamma(\ln c_i - \ln \bar{c}_i)]\, s_{i,t-h} \qquad \text{for } i \leq N_t, \tag{2.6}$$

where N_t is the index of BPT at time t and t does not belong to $U(N_t)$, where $U(N_t)$ is the setup phase to build BPT and $s_{i,t}$ is the capacity shares of the technique i at time t. Note that the capacity share of the techniques changes inversely in proportion to the percentage deviations of their unit costs from the economy-wide average unit cost $\bar{c}$ where

$$\ln \bar{c}_t = \sum_{i \leq N_t} s_{i,t} \ln c_i . \tag{2.7}$$

Franke (2000) has discussed three important implications of the Iwai model, which extended the Schumpeterian dynamics of interaction between technological innovation and diffusion. The first implication deals with the speed of diffusion γ. The Iwai model showed that this parameter γ can be approximately expressed as

$$\gamma \sim \beta_s \rho w / k,$$

where β_s is the constant propensity to save out of profits, w is the share of wages in national income, k is the capital–output ratio, and ρ is a constant speed of adjustment. Thus the speed of diffusion depends on the firm's responsiveness to the profitability of alternative technologies, i.e., higher capital–output ratios would tend to lower the speed of diffusion, while higher profitability induces greater speed of diffusion.

The second implication is that the differential equations of the mode describe an evolutionary process that tends to steer the economy's state of technology toward an equilibrium in which all firms use the most efficient production function but the function of innovation lies precisely in upsetting this equilibrating tendency. The Iwai model derives an equilibrium trajectory Φ^e where $\Phi = \{s_{i,t} : i \in N, t = 0, h, 2h \ldots\}$ and Φ^e denotes a balanced growth path. This equilibrium Φ^e corresponds to what is known as a traveling wave or a wave train in the literature on partial differential equations. The significance of the equilibrium concept of a wave train depends on its global stability. Finally, Franke has shown by a series of simulations that across a wide range of parameter scenarios the average wave lengths underlying the long oscillation in the growth rates of average productivity are found to be closely related to the lifetime of techniques in the underlying economy. One important extension of this Iwai–Franke model lies in an endogenous transformation of the model, where profitability can be directly related to the diffusion process.

C. A Cournot–Nash Model of Spillover

Innovations in investment and knowledge capital tend to reduce unit costs of firms and industries and thereby build competitive pressure across firms. They also generate significant external economies and spillover effects. These spillover effects occur at several levels. First, we have the new growth theory which introduced several endogenous factors, which challenged the basic assumption of the Solow model that technology alone determines the long-run growth of an economy and that this technology is completely exogenous in the sense that it is unaffected by profits and market opportunities. Endogenous growth theory emphasizes the spillover and externality effects through technology diffusion. One simple form of endogenous growth model is the AK model:

$$Y(t) = A(k)K(t)$$

with real income (Y) as a linear function of $K(t)$ which is a composite measure of the combined stock of human, physical, and knowledge capital. Here $A(k)$ represents endogenous technical change depending on per capita K. Different economies will have different $A(k)$ values depending on their pattern of knowledge creation and technological diffusion and spillover. A basic premise of endogenous growth theory is that technology or innovation in knowledge capital is in part a public good. Private good market incentives and profitability both influence its development, whereas the public good property stresses the point that not all cost economies from knowledge creation and accumulation can be internalized. Diffusion and spillover and interfirm interdependence help spread the innovation. Endogenous growth theory assumes that an economy automatically benefits from its investments in new knowledge (Lucas 1993; Romer 1990) because knowledge has a public good aspect that can be used by an entire economy, leading to innovation and growth. Braunerhjelm et al. (2010) use the knowledge spillover theory of entrepreneurship derived from the basic spillover theory of knowledge to develop a theoretical model in which the transformation of knowledge into economic growth depends on how knowledge diffuses through both incumbent and entrepreneurial activity. The entrepreneur is the missing link in converting knowledge into economically relevant knowledge. Based on OECD data from 1981 to 2002 they show that entrepreneurship Granger-causes economic growth and that this effect increased in the 1990s and later as the knowledge economy began to grow.

Endogenous growth theory suggests that active government policies can definitely affect the long-run rate of economic growth by impacting the accumulation of composite capital K in the AK model.

The second aspect of knowledge spillover theory arises through the effort and investment allocated to R&D and the diffusion knowledge through software development and other linkages provided by the new information technology. Lucas has stressed the concept of learning spillover technology as an important feature of endogenous technology. This spillover is the source of rapid productivity growth and cost economies due to increasing returns to scale. To the question why does net capital flow from rich to poor countries on a large scale, the answer provided by the Lucas model is that the spillover effect is very small in poor countries. Lucas introduced several new dimensions of endogenous technology as innovations. He pointed out that the Asian growth miracles, e.g., high sustained growth in five Asian countries, Japan, South Korea, Hong Kong, Singapore, and Taiwan, over 1965–2005, cannot simply be explained by physical capital accumulation alone. One has to introduce the dynamic role of human and knowledge capital.

A third aspect of knowledge spillover theory is its impact on the globalization of markets in high-tech goods and services. Competitive pressures have increased and the pressure of significant increasing returns to scale in new high-tech industries disrupted the guiding principles of Walrasian competitive equilibria. The new market structure that has evolved in these new industries such as telecommunications, computers, microelectronics, and bioengineering is dominated by large firms which enjoy large economies of scale due to innovation-based investment. This type of market structure has been called by D'Aveni (1994) "hypercompetition," which diverges from a competitive market structure in several ways. First of all, it is driven by knowledge capital and innovative investments. Second, it augments the various strategies of nonprice competition. Mergers and acquisitions, cooperative ventures in R&D networks have led to decline in unit costs and prices, resulting in Cournot–Nash-type equilibrium solutions.

Unlike Walrasian adjustment, the Cournot–Nash framework emphasizes the game-theoretic interdependence of duopoly or oligopoly firms. Their mutual reaction functions describe the dynamic process of adjustment and several types of equilibria are possible. Three types of equilibria are most important. One is the dominant firm model often discussed in limit pricing theory. There is one dominate firm but it cannot set monopoly price because of the potential threat by firms on the competitive fringe. This may also lead to a leader–follower model, where the dominant firm is the leader and the rest are followers. Secondly, firms may compete in an oligopoly framework, where innovations through R&D investments generate spillover effects that cannot be internalized by individual firms. In this case firms may form a cartel so as to internalize spillover effects or act independently in a Cournot–Nash competition. Thirdly, there is the hypercompetitive market framework, where dynamically efficient firms follow the growth frontier and sustain it, whereas inefficient firms fail to compete and exit the market. We would briefly discuss these noncompetitive selection mechanisms for industry growth.

A dominant firm in the context of a limit pricing model may be a leader with a large market share, where the follower's reaction functions to the leader's strategy are already incorporated in the leader's optimal output and pricing strategies. However a dominant firm cannot adopt a price monopoly strategy due to the possibility of new entry. Cost-reducing innovation strategy therefore offers a long-run optimal strategy for the dominant firm.

To consider this type of cost-reducing innovation strategy we consider a dynamic cost-reducing model of innovation capital k where the objective is to maximize the discounted profit stream:

$$\max_u \pi(k_0) = \int_0^\infty e^{-\rho t}(r(k) - c(u))\mathrm{d}t$$

$$\text{subject to } \dot{k} = u - \delta k,\ k(0) = k_0 > 0,$$

where u is investment. The revenue $r(k)$ and cost function $c(u)$ are assumed to be concave, i.e.,

$$r(k) = ak - bk^2$$

$$c(u) = c_1 u - c_2 u^2,$$

where all parameters a, b, c_1, c_2 are positive. The cost function exhibits economies of scale, i.e., unit cost declines as investment rises.

On using the Hamiltonian function

$$H = ak - bk^2 c_1 u + c_2 u^2 + q(u - \delta k)$$

we derive the adjoint equation

$$\dot{q} = \frac{\mathrm{d}q}{\mathrm{d}t} = \rho q - \frac{\partial H}{\partial k} = (p + \delta)qa + 2bk.$$

This yields

$$\dot{u} = (p + \delta)u + \frac{(p + \delta)c_1 - a}{2c_2} - \frac{bk}{c_2}$$
$$\dot{k} = u - \delta.$$

The characteristic equation has two roots:

$$\lambda_1, \lambda_2 = \frac{1}{2}\left[\rho \pm \left\{(\rho + 2\delta)^2 - \frac{4b}{c_2}\right\}^2\right].$$

It follows that λ_1 is negative, while λ_2 is positive. Hence we have to consider only the stable root λ_1 where the growth path of $k(t)$ converges to the steady-state equilibrium values $\bar{k}$ and $\bar{u}$:

$$\bar{k} = \left(\frac{1}{2}\right)\frac{c_1(\rho + \delta) - a}{b - c_2\delta(\rho + \delta)}$$
$$\bar{u} = \delta\bar{k}.$$

The equilibrium is a saddle point if and only if

$$\delta c_2(\rho + \delta) < b.$$

The existence of a stable manifold converging to the saddle point equilibrium for the dominant firm shows a viable strategy for the innovation investment.

The strategic interaction between the dominant firm and the competitive fringe can be recast as a dynamic limit pricing model where the dominant firm sets the price and the fringe enjoys lower production costs due to newer technology.

Recently Cellini and Lambertini (2009) have formulated a dynamic oligopoly model, where the firms may undertake independent research ventures or form a cartel for cost-reducing R&D investments. We consider here a duopoly version, where $q_i(t)$ are the outputs ($i = 1, 2$) and the market demand and unit cost functions are

$$p(t) = A - q_1(t) - q_2(t)$$

$$\frac{\dot{c}_i(t)}{c_i(t)} = -k_i(t) - \beta k_j(t) + \delta, i \neq j,$$

where dot is time derivative, $k_i(t)$ is the R&D effort of firm i, and δ is the constant rate of depreciation. The parameter β with $0 < \beta < 1$ denotes the positive technological spillover that firm i receives from firm j. When each firm behaves independently, the cost of setting up a single R&D laboratory is assumed to be of the form

$$G_i(k_i(t)) = b(k_i(t))^2, b > 0.$$

On applying Pontryagin's maximum principle and assuming the case of independent R&D ventures, each firm maximizes a discounted profit function:

$$\max_{q_1, q_2} \pi_i(t) = \int_0^\infty e^{\rho t} \left[(A - q_i(t) - q_j(t) - c_i)q_i(t) - b(k_i(t))^2\right] \mathrm{d}t$$

$$\text{subject to } \frac{\dot{c}_i(t)}{c_i(t)} = k_i(t) - \beta k_j(t) + \delta; i \neq j; i = 1, 2.$$

On using the present value Hamiltonian, one can derive the optimal conditions denoted by asterisks:

$$q_i^* = \left(\frac{1}{2}\right)(A - q_i(t) - c_i(t))$$

$$k_i^* = \frac{-\lambda_{ij}(t)c_(t) - \beta\lambda_{ij}(t)c_j(t)}{2b},$$

where $\lambda_{ij}(t) = \mu_{ij}(t)e^{-\rho t}$ is the present value costate variable for the control variable $c_i(t)$. These two equations describe the standard Cournot–Nash reaction functions. If we satisfy the condition $\delta\rho \leq \frac{A^2(1+\beta)}{24b}$, then there is a saddle point equilibrium with steady-state values

$$\bar{c} = \frac{A(1+\beta) - \left\{(1+\beta)\left[A^2(1+\beta) - 24b\beta\delta\right]\right\}^{1/2}}{2(1+\beta)}$$

$$\bar{k} = \delta(1+\beta)^{-1},$$

where $c_1(t) = c_2(t) = c(t)$ (case of symmetry assumed) clearly $\partial k/\partial\beta < 0$, i.e., an increase in the spillover effect β leads to a decrease in the steady-state equilibrium level of k. This suggest the need for remedial public sector policies.

In case the firms form a cartel in the R&D style the firms choose output levels noncooperatively while maximizing joint profits. In this case the steady-state levels of c and k are

$$\begin{aligned}\bar{c} &= [2(1+\beta)]^{-1}[A^2(1+\beta)^2 - 24b\rho\delta]^{1/2}\\ \bar{k} &= \delta(1+\beta)^{-1}.\end{aligned}$$

The steady-state levels of R&D effort $\bar{k}$ are the same in this case, but the level of unit cost in the case of Cartelization is lower. The extent of consumer surplus (CS) in the steady state however is much lower for the case of cartel compared to the case of independent ventures, since we have

$$\begin{aligned}\text{CS(cartel)} &= [18(1+\beta)^2]^{-1}\left[A(1+\beta) + \{A^2(1+\beta)^2 \right.\\ &\qquad \left. + 24b\delta\rho\}^{1/2}\right]\\ \text{CS(independent ventures)} &= [18(1+\beta)]^{-1}\left[A(1+\beta)^{1/2}\right.\\ &\qquad \left. + A^2(1+\beta)^2 24b\delta\rho^{1/2}\right]^2.\end{aligned}$$

Note also that the steady-state unit cost is much higher for the case of independent ventures and hence the cooperative R&D in case of cartelization would help increase R&D investment more than the case of independent ventures.

Modern firms in the information technology sector today have several economic incentives to cooperate and combine R&D efforts. First of all, the technology of the new innovation process is becoming increasingly complex and the initial cost of development, a fixed cost, is becoming very large. Second, there is increasingly the possibility that the competitors may copy the new technology, e.g., software technology. Third, collusion and cooperation in the R&D phase may help the innovating firms to internalize a large portion of the spillover effects and thereby reduce unit costs and gain larger market shares. By now the governments in most industrial countries have recognized this need. For example, the European Commission allowed in March 1985 a 13-year block exemption under Article 85(3) of the Treaty of Rome to all firms forming joint ventures or cartel in R&D.

d'Aspremont and Jacquemin (1988) analyzed the collaborative R&D situations and compared them in some detail with noncooperative R&D levels. The model of Cellini and Lambertini is only a dynamic extension of the A&J model. Their major conclusion is that optimal cooperative R&D levels exceed those of noncooperative R&D, whenever technological spillover is relatively large (i.e., above 50 %) while the opposite holds for small spillover below 50 %. These results imply that the antitrust authorities should encourage the formation of joint ventures in R&D with a sharing of all information but without allowing collusion in the product market.

D. Demand-Induced Model of Innovation

Modern innovations have two major endogenous impacts. One is in terms of significant market expansion both locally and globally. Secondly, there have occurred significant economies of scale in demand fostered by economies of scale in supply. The competitive advantage (CA) principle emphasized by Baumol (2002); Porter (1990), and others has discussed this cumulative process of demand-induced growth of endogenous innovation.

Modern economies have undergone a fundamental transformation today due to the widespread use of computers and communication technology. The shift from traditional large-scale material manufacturing to the use of new technology and software networks has introduced three profound changes in industrial structure all over the world. Gone are the days of diminishing returns industries. The increasing returns and scale economies have dominated the new technology increasingly using knowledge and innovations in software and networking methods. This has resulted in decreasing unit production costs and increasing productivity. Modern technology involves high fixed cost for the initial innovation but very low or negligible marginal cost, e.g., iPod and iPhone. Secondly, this technology generates high network effects which involve increasing value of products as more and more users use or adopt the product or the process, e.g., Windows 7. This is sometimes called scale economies in demand. Finally, new technology frequently involves high switching costs, so that the users once locked in find it difficult to switch to alternative products. All these characteristics of modern technology involve two major impacts on the industrial structure. One is that the competitive paradigm of the market structure no longer holds. Hence various types of noncompetitive structures have to be analyzed. The dynamic model formulated by Spence (1984) and others like the dynamic limit pricing model have to be analyzed in the new paradigm. These models discuss the welfare implications of declining cost industries subject to noncompetitive structures. A second type of model considers industry growth through the entry dynamics. Sengupta (2007) has discussed several types of dynamic entry models and market evolution. Most of these models view entry either as entering into an existing market or as an increase in market share of an existing industry through unit cost reduction due to new technology or innovation.

The Spence model considers markets where firms compete over time by investing resources for reducing unit costs. In many instances their strategies take the form of developing new products at cheaper costs. Cost-reducing expenditures like R&D investments (e.g., research for new drugs) are largely fixed costs with very little marginal costs. As a result the market structures are likely to be concentrated and imperfectly competitive. Cournot–Nash-type equilibria are more appropriate in this environment. The scale economies and product differentiation are two important characteristics of this environment. Two important economic issues arise here. One is the spillover or externality effect of R&D investment and dynamic innovation. These spillover benefits are internally appropriable. Hence firms have to devise alternative methods like cooperative ventures in R&D to share the benefits. Secondly, while spillovers reduce the incentives for cost reduction through innovation, they can also reduce the costs at the industry level of achieving a given level of cost reduction. However, the incentives can be restored through subsidies. Many high growth countries in Southeast Asia like China, Taiwan, and Singapore have adopted these strategies through direct state subsidies to R&D innovations.

In the dynamic entry model the efficient firms tend to increase their market share through cost-reducing strategies. A typical model here takes the form

$$\frac{\dot{y}_j}{y_j} = a(\bar{c} - c_j), a > 0 \tag{2.8}$$

$$c_j = f(I_j), \frac{\partial c_j}{\partial I_j} < 0,$$

where dot denotes time derivative. Here c_j and y_j are unit cost and output of the efficient firm j which invests I_j to reduce unit costs and $\bar{c}$ denotes average costs of other firms in the industry. When c_j falls or $\bar{c}$ rises, the efficient firm increases its output resulting in an increase in market share. When investment I_j follows its optimal expansion path, $\dot{c}_j$ falls and therefore $\frac{\dot{y}_j}{y_j}$ increases. On replacing y_j by the market share of the efficient firm, this relation (2.8) can be directly used for empirical testing.

An alternative framework for analyzing the cost-reducing aspect of R&D investment is through a Pareto efficiency model applied to n firms in an industry. This may be done through a sequence of linear programming (LP) models also known as DEA. Two types of formulations may considered here. One emphasizes the cost-reducing impact of R&D inputs. This may be related to the learning-by-doing implications of knowledge capital. Secondly, the impact on output growth through R&D investment may be formulated as a growth efficiency model. Here a distinction is drawn between the level and growth efficiency, where the former specifies a static production frontier, while the latter a dynamic frontier. Denote unit (or average) cost of any firm j by $\frac{c_j}{y_j}$ where total cost c_j excludes R&D costs denoted here by r_j instead of I_j. Then we set up the Pareto efficiency model (DEA) with radial efficiency scores θ:

$$\min \theta, \text{ subject to}$$

$$\sum_{j=1} nc_j\lambda_j \leq \theta c_h \qquad \sum_{j=1} nr_j\lambda_j \leq r_h$$

$$\sum_{j=1} nr_j^2\lambda_j = r_h^2 \qquad \sum_{j=1} ny_j\lambda_j \geq y_h$$

$$\sum_{j=1} n\lambda_j = 1; \lambda_j \geq 0; j, h \in I_n = \{1, 2, \ldots, n\}.$$

On using the dual variables $\beta_0, \beta_1, \beta_2, \beta_3$, and α and solving the LP model above for a firm j which is Pareto efficient, we obtain the optimal values $\theta^* = 1.0$ and zero for all slack variables with the following average cost frontier:

$$c_j^* = \beta_0^* - \beta_2^* r_j + \beta_3^* r_j^2 + \alpha^* y_j,$$

where asterisk denotes optimal values and $\beta_0^* = 1.0$ if $\theta^* = 0$. Thus if R&D spending r_j rises, average cost c_j falls for the efficient firm if $2\beta_3^* r_j < \beta_2^*$. If we replace r_j by cumulative R&D R_j as in learning-by-doing models where R_j is cumulative experience, then the AC frontier becomes

$$c_j^* = \beta_0^* - \beta_2^* R_j + \beta_3^* R_j^2 + \alpha^* y_j.$$

So long as the coefficient β_3^* is positive, r_j may also be optimally chosen as r_j^*, if we extend the objective function in the LP model as min $(\theta + r)$ and replace r_h by r. In this case the optimal value of R&D spending r^* is

$$r^* = (2\beta_3^*)^{-1}(1 + \beta_2^*).$$

A similar result follows when we use the cumulative R&D spending R_j or R here. This framework can be easily extended to the case of multiple inputs and outputs.

Now consider a Pareto model of growth efficiency frontier. Consider a firm j producing a single (composite) output y_j with m inputs x_{ij} by means of a log-linear production function:

$$y_j = \beta_0 \prod_{i=1}^{m} e^{B_i} x_{ij}^{\beta_i}, \; j = 1, 2, \ldots, n,$$

where the e^{B_i} denotes the industry effect as a proxy for the share of total industry R&D. On taking logs and time derivatives of both sides we can easily derive the production frontier:

$$Y_j \leq \sum_{i=0}^{m} b_j X_{ij} + \sum_{i=1}^{m} \phi_i \hat{X}_i$$

$$\text{when } b_i = \beta_i, b_0 = \frac{\dot{\beta}_0}{\beta_0 X_{0j}}, \; j = 1, 2, \ldots, n$$

$$e^{B_i} = \phi_i \hat{X}_i, \; X_{ij} = \frac{\dot{x}_{ij}}{x_{ij}}, \; Y_j = \frac{\dot{y}_j}{y_j}$$

$$\text{and } \hat{X}_i = \frac{\sum_{j=1}^{n} \dot{x}_{ij}}{\sum_{j=1}^{n} x_{ij}}$$

and dot denotes time derivative. Note that b_0 here denotes technical progress in the sense of Solow residual and ϕ_i denotes the input specific industry efficiency parameter. In the Pareto efficiency or DEA model we test the relative growth efficiency of each firm k in an industry of n firms by the LP model

$$\min C_k = \sum_{i=0}^{m} (b_i X_{ik} + \phi_i \hat{X}_i)$$

$$\text{subject to } \sum_{i=0}^{m} (b_i X_{ij} + \phi_i \hat{X}_i) \geq Y_j,\ j = 1, 2, \ldots, n$$

$$b_0 \text{ free in sign; } b_1, b_2, \ldots, b_n \geq 0;\ \phi_i \geq 0.$$

Denoting the optimal solutions by asterisks one can derive as before the following results: a firm k is growth efficient if

$$Y_k = b_0^* + \sum_{i=0}^{m} (b_i^* X_{ik} + \phi_i^* \hat{X}_i).$$

In case the equality sign changes to the "greater than" sign $>$, then the kth firm is not growth efficient, since the observed output growth Y_k is less than that of the optimal output. This growth efficiency model can be used to compute two subsets of firms: one growth efficient, the other not so. Clearly the industry growth would be dominated by the growth-efficient firms. Technology and innovations would play a catalytic role here. Also we can compare the level efficiency here with the growth efficiency.

Innovations have two dynamic characteristics. One is their impact on production costs and economic efficiency. This occurs through upward shifts in the production frontier. Frequently this involves a race in R&D investments among competing firms. It also leads to quality improvements of existing goods and services. For example, a pharmaceutical firm develops an improved drug through R&D investment over several years. It then becomes the new leader, the winner of the R&D race. It raises the price exactly to the extent of the quality improvement. At this price the leading firm becomes a monopoly producer, because infinitesimal price reductions allow it to take over the market. Economic efficiency increases for the industry as a whole due to what Schumpeter called "creative destruction," i.e., old processes or products cannot survive the new competition and die out. A second dynamic aspect of innovation is the process of routinization of innovations in oligopolistic competition and the spread of incremental innovations. The latter involves industry-wide transmission of new technology and cumulative multiplier effects. Baumol (2002) has considered this process as the dynamic engine of unprecedented capitalist growth in modern times. We would discuss in this section two important models developed by Baumol. One is the technology consortium model, where he characterizes the cost of nonmembership. The second model develops optimal rules for recouping innovation outlays that involve large fixed costs. Technology knowledge by innovation is itself a kind of capital good that can be accumulated through R&D and other knowledge-creation activities. It goes far beyond the Schumpeterian notion of creative destruction.

The R&D race model considers an industry consisting of more or less homogeneous firms engaged in R&D competition. The instantaneous net profit of a representative incumbent firm is a function of the number of firms n in the industry and of an R&D parameter u so that $\pi = \pi(n, u)$. The R&D parameter u is the effort made by the firm in product innovation at time t. Given n, the incumbent firm maximizes current profits to obtain optimal R&D effort:

$$u(n) = \max_{u} \pi(n, u)$$

assuming the profit function to be concave in u. For the long run each incumbent firm chooses the time path of R&D that maximizes the present value of staying in the industry indefinitely, i.e.,

$$v_0 = \int_0^\infty e^{-rt} \pi(n, u) \mathrm{d}t,$$

where r is the real discount rate assumed to be a positive constant. If one firm innovates successfully, it will be a leader during the subsequent time period until another firm wins the R&D race. Any winner earns monopoly profits. The expected monopoly profits for the successful winner depend on the expected monopoly surplus due to higher price equaling quality improvement and the probability that the firm innovates successfully.

The winner of the R&D race may reap another important benefit, i.e., through innovative R&D it may augment its productivity significantly. In that case the exploitation of scale economies may generate a higher market share for the leading firm. In this case the leading firm may play the role of a dominant firm; the other firms in the fringe are then the followers in a Bertrand game. The dominant firm may attempt to maintain its dominance in market share through innovations based on up-to-date R&D and also prevent potential entry.

Baumol (2002) has stressed the distinction between routinized vs. nonroutinized innovations, implying that the latter affects industry growth in a cumulative fashion. Also it intensifies the impact of the competitive advantage principle in a global fashion. Independent nonroutinized innovations can be viewed as dynamic shocks to the static equilibria of the Walrasian competitive paradigm. They may involve new processes, new products, or new markets. Baumol (2002) has discussed in some detail three growth creating properties of nonroutinized innovations as follows:

1. The cumulative character of many independent innovations, which not only replace old technology but also create new technical knowledge. The spillover effect is thus enhanced and other firms can utilize such spillover to reduce their unit costs and prices. Many successful NICs in Taiwan, China, and Korea

have used deliberate state policies to intensify the transmission of this spillover process.

2. The public good property of such innovations, which imply economies of scope in the generation of this new technological knowledge. This generates the adverse effect of reducing the optimal levels of innovation investment. Appropriate public policy is therefore needed here to correct the imbalance.
3. This type of innovation generates accelerator effects of induced investment, where the innovating sector's output and investment growth help other sectors grow through forward and backward linkage. There is considerable scope of state action in this framework. In many successful NICs of Southeast Asia, industrial parks, hubs, export zones, and technology consortia have been deliberately sponsored by the state as a sharing center of new knowledge about the latest technology and software.

We may now add a few comments on the CA theory in global trade and its diffusion, which emphasize industry growth due to the innovation process reducing unit costs and prices of new technology-intensive products.

Competitive advantage (CA) principle has two basic features. One is that the firm with CA earns a higher rate of economic profits than the average rate of economic profit earned by other firms in the market. Thus to assess if the technology firm Sun has a CA in its core business of designing and selling high-technology company servers, we would compare Sun's profitability in this business to the profitability of such firms as IBM and HP that also sell enterprise servers. The second feature of CA is higher competitiveness of firms with CA. In international trade this is revealed through relative cost advantage of successful firms dominating the international market. Growth in modern technology and knowledge diffusion through the information and communication technology have expanded the market structure to global levels and CA can be measured in this framework through (1) technological competitiveness T/T_w, (2) price competitiveness P/P_w, and (3) capacity utilization C. Here T and P denote technology development index and price per domestic good. The subscript w in T and P denotes the world levels. Fagenberg (1988) has measured the economic effect of CA in international trade through increase in export share through a multiplicative functional form as

$$S = AC^{v}\left(\frac{T}{T_w}\right)^{e}\left(\frac{P}{P_w}\right)^{-a},$$

where A, v, e, a are positive constants. On differentiating with respect to time (denoted by a dot over the variable) we obtain

$$\frac{\dot{S}}{S} = v\left(\frac{\dot{C}}{C}\right) + e\left(\frac{\dot{T}}{T} - \frac{\dot{T}_w}{T_w}\right) - a\left(\frac{\dot{P}}{P} - \frac{\dot{P}_w}{P_w}\right).$$

He further measured capacity advantage in terms of the ability to deliver at cheaper price. This improved ability is assumed to depend on three factors: (a) the growth in technological capability and knowledge diffusion of technology along the world innovation frontier $\dot{Q}/Q$, (b) the growth in physical capital and infrastructure $\dot{K}/K$, and (c) the growth in demand $\dot{D}/D$.

$$\frac{\dot{C}}{C} = \alpha_1\left(\frac{\dot{Q}}{Q}\right) + \alpha_2\left(\frac{\dot{K}}{K}\right) - \alpha_3\left(\frac{\dot{D}}{D}\right),$$

where α_1, α_2, α_3 are positive constants. He assumed knowledge diffusion to follow a logistic curve:

$$\frac{\dot{Q}}{Q} = \beta - \beta\left(\frac{Q}{Q^*}\right),$$

where β is a positive constant and Q/Q^* is the ratio between the country's (or firm's) own technological development and that of the countries on the world innovation frontier. On combining the equations above we obtain the final equation for CA of firms in international trade.

$$\frac{\dot{S}}{S} = v\alpha_1\beta - v\alpha_1\beta\left(\frac{Q}{Q*}\right) + v\alpha_2\left(\frac{\dot{K}}{K}\right) - v\alpha_3\left(\frac{\dot{D}}{D}\right) + e\left(\frac{\dot{T}}{T} - \frac{\dot{T}_w}{T_w}\right) - a\left(\frac{\dot{P}}{P} - \frac{\dot{P}_w}{P_w}\right).$$

This model was empirically tested on pooled cross country and time series data for the period 1960–1983 covering 15 industrial countries (mostly OECD countries) and the results show that the main factors influencing differences in international competitiveness and growth across countries measured by export shares are technological competitiveness and the dynamic ability to compete in satisfying world demand measured by efficiency of capacity utilization. Recent experiences of rapidly growing economies of Southeast Asia have exhibited the dynamic role of technological and cost competitiveness in achieving high export performance in world markets.

Recently Porter (1990) made a comparative study of the sources of growth of rapidly growing countries of the world and found that the only meaningful concept of competitiveness through CA at the national level is national productivity which is measured by the firms moving along the innovation frontier. Three basic points are central to competitive advantage, e.g., (1) scale economies, (2) technological change, and (3) quality improvements and new product innovations.

In global competition firms from any nation can gain scale economies by selling worldwide. Comparative advantage theory in trade helps explain in part the specialization in specific commodities for the advanced industrial countries. Thus the Italian firms reaped the economies of scale in appliances, German firms in chemicals, Swedish firms in mining equipment, and the Swiss firms in textile machinery. The second point in competitive advantage model is recently stressed in the "technology gap" theories in which nations will export in industries in which their firms gain a lead in technology. Exports will then fall as technology diffuses over time and the spillover effect spreads and the gap closes. Finally, the spearheading of new products and quality improvements has been intensified in world competition through the spread of multinational corporations. Their prominence in world trade means that trade is no longer the only important form of international competition. Recent empirical suggests that a significant portion of world trade is between subsidiaries of multinationals. National success in an industry increasingly implies that the nation is the home base for leading multinationals in the industry, not just for domestic firms that export.

Porter's theory of competitive advantage (CA) of nations comprises several new features, e.g., (1) it moves beyond the comparative advantage theory of international trade which is restricted to limited types of factor-based advantages, (2) it extends the Schumpeterian model of innovation by asking why do some firms, based in some nations, innovate more than others, (3) it explains how firms gain CA from changing the constraints, i.e., by improving the equality of factors, raising productivity, and creating new products, and (4) it emphasizes the managerial perspective in creating competitive advantage.

To investigate why countries gain competitive advantage in particular industries, Porter studied ten countries, Denmark, Germany, Italy, Japan, South Korea, Singapore, Sweden, Switzerland, the UK, and the USA, over a 4-year (1985–1988) study. One has to note that this list of countries includes four Southeast Asian countries, which are very important among the NICs in Asia which have achieved rapid growth rates in the last three decades. It is instructive to analyze the sources of rapid growth in these countries which have successfully excelled in world competition in modern technology-intensive products.

In global markets today competitive efficiency holds the key to economic success. Porter's study of ten industrially successful countries reached four important conclusions. First, sustained productivity growth at the industry level requires that an economy continually upgrade itself. A country's growing firms must also develop the capability to compete in more new and more sophisticated industry segments. At the same time an upgrading economy is one that develops the capability of competitive success in entirely new and modern industries.

Secondly, firms gain competitive advantage from conceiving new ways to conduct activities, employing new technologies or different inputs. Thus Makita in Japan emerged as a leading competitor in power tools because it was the first to employ new and less expensive materials from making tool parts. Gaining competitive advantage requires that a firm's value chain is managed as a system rather than as a collection of separate parts. A good example is in appliances, where Italian firms transformed the channels of distribution to become world leaders in the 1970s. Likewise Japan in cameras. Firms generate competitive advantage by discovering new and better ways to compete in an industry. Porter identified five sources of innovations that shift CA as follows:

1. New technologies
2. New buyer needs
3. Emergence of a new industry segment

4. Shifting input costs such as labor and knowledge capital
5. Liberalization of government regulations

The last source has played a most dynamic role in the wave of economic reforms introduced in China, Taiwan, and South Korea, which has achieved a very high growth rates and then sustained it over the last three decades.

Thirdly, the CA principle is basically dynamic and hence it thrives under competitive international trade. Trade allows a country to raise its productivity by specializing in those industries in which its firms are relatively more efficient. This allows exports to grow with multiplier effects in the domestic sectors through linkages. For new technology transfer the countries specializing in the efficient sectors may gain early mover advantages such as being the first to reap economies of scale, reducing costs through cumulative learning and R&D knowledge spillover.

Finally, one must note the dynamic role of sustainability. CA is sustained by constant improvement and upgrading. This is precisely what Japanese automakers have done. They initially penetrated foreign markets with inexpensive compact cars of adequate quality and competed on the basis of lower labor costs. Then they became innovations in process technology. Sustaining competitive advantage requires change and innovation. It demands that a company exploit rather than ignore industry trends. In many situations an innovation firm has to destroy old advantages to create new higher-order ones. This is what Schumpeter called "creative destruction." For example, South Korea's shipbuilding firms did not become international leaders until they aggressively expanded the scale and scope of new changes in technology.

Two Asian economies, Taiwan and South Korea, have to be mentioned as special examples of success in rapid growth, where the competitive advantage principle has been applied to a significant degree. The scale and scope of application of this principle has been widespread across the new industries competing intensely in international trade.

Korean Case

Three basic features about Korean growth have been emphasized by Porter in his empirical study. First, Korea has made major investments in factor creation, well beyond those of most other successful Asian NICs. This is a major reason why it has been able to upgrade its economy and compete in international markets. It has a high level of literacy and a high average level of education with universal education into the high school level. A survey performed by the Economic Planning Board in 1987 found that 84.5 % of Korean parents wanted to provide their children with a college level education. The university system is extensive and particularly aggressive investments have been made in engineering. Korean companies above a certain size are required by law to provide training for their employees. It is typical for a large Korean group of companies to invest $25 to 30 million in training facilities alone. Major Korean companies also invest heavily to upgrade their technical capability compared to companies from other developing countries. High rates of R&D to sales ratio are typical for most modern companies. Korean firms are unique among firms from other NICs in their commitment to developing their own product models and to investing in the up-to-date process technology.

Porter has stressed several important features of Korean companies, which utilize the CA principle in remarkable ways. First, the most unique feature of almost all modern Korean companies is their utmost willingness to take risk. Companies rush into industries and make huge investments in plant and equipment in advance of any substantial orders. In shipbuilding, for example, Hyundai and Daewoo built huge shipyards before the orders arrived to fill them. In videotape industry all four of their leading firms (e.g., Sachan, SKC, Lucky–Goldstar, and Kolon) have more than doubled installed capacity in 1987–1990, despite having already achieved about 25 % of the world market.

Second, Korean companies in high-tech fields face fierce competition in domestic fields, e.g., in automobiles, computer semiconductors, shipbuilding, steel, fabrics, TV sets, and memory chips. This domestic competition creates continued pressure to invest, improve productivity, and introduce new products. The Korean government has played a dynamic productive role in this competitive process. One of the unique historical strengths of Korean government policy has been its capacity to adjust and evolve and thereby help the process of industry growth.

Another unique feature of the Korean industry is the importance of the large groups called the *chaebol*. Companies such as Hyundai, Samsung, and Lucky–Goldstar contribute close to 40 % of world exports by some estimates. The *chaebol* have been favored and heavily supported by government. That is why they are able to take larger risks than in other Asian NICs.

Finally, the Korean economy is largely innovation driven. Three aspects of this innovation drive have to be made. One is that the more advanced firms in this economy develop increasingly sophisticated service needs in engineering, testing, and marketing. Secondly, the companies not only import advanced technology from other nations but create them. Learning by doing is actively followed by the heavy emphasis on human resources, skills, and R&D by both government and private firms. Thirdly, a new form of Schumpeterian "creative destruction" strategy has been consistently adopted by the progressive Korean firms. Thus selective cost disadvantages in design and technology have helped stimulate new innovations that advance product and process technology. Industry clusters and research centers augmented the industry capacity to innovate more new industries and their ancillaries.

Taiwan Model

Taiwan's rapid industry growth has two important differences from the Korean model. First, it has emphasized small and medium industries much more than the large ones. As a result, the resulting income distribution has been more equitable. The so-called Kuznets hypothesis which asserts a close positive correlation of economic growth with inequality of income distribution has been found not to hold for Taiwan. Secondly, much of rapid growth in China and Hong Kong over the last three decades has been contributed by Taiwan and its investment in new processes and innovations.

Three aspects of the Taiwan model of growth deserve special mention: (1) impressive record of the IT (information technology) sector, (2) emphasis on decentralized industry development, and (3) sound macroeconomic policy emphasizing economic efficiency in governance.

Taiwan's contemporary knowledge-based economy has revealed more remarkable growth of the IT sector than China and other Asian NICs. From 1995 to 1999, Taiwan's IT industry ranked third in the world after the USA and Japan. Taiwan's strong leadership in R&D and other investment in the IT sector started in 1982, when the value of exports of IT products was only $106 million in US dollars, but by 1985 these exports climbed to $1.22 billion, representing about 1 % of world market share. The overall R&D intensity rose from 1.78 in 1995 to more than 2.90 in 2008. The World Economic Forum (2004) has computed a growth competitiveness index (GCI) based on three components: infrastructure development, efficiency of public institutions, and the use of best practice technology. Here Taiwan's record of performance in the IT sector is most impressive. In terms of average number of annual US patents per million people, the top rankings in the world in 2004 were 1 for the USA, 2 for Japan, and 3 for Taiwan. The numbers of patents were 301.48 (USA), 273.40 (Japan), and 241.38 (Taiwan). Singapore ranks 10 and South Korea 14.

Traditional technology is usually subject to diminishing returns. Modern technology however is different. It involves improvement in the productivity of knowledge and R&D investment viewed as "knowledge capital." This capital input is complementary to all other inputs associated with the production function. An economy characterized by this new technology is often called "the new knowledge economy" and it has four fundamental characteristics: accumulating knowledge capital through R&D, improving competitive efficiency, expanding export markets through global trade, and increased collaboration utilizing the external benefits of new technology. Knowledge capital may take several forms, e.g., (1) software development, (2) new designs and blueprints, (3) R&D investments for new products involving "creative destruction" of old process, and (4) skill development through learning by doing. The successful NICs in Asia have developed this new knowledge capital and Taiwan has evidenced a remarkable record performance over the last three decades.

Both China and Taiwan have made consistent attempts to follow the paradigm of competitive market capitalism, where private industries compete for efficiency and growth. An important element of China's and also Taiwan's growth experience is its spread across regions and sectors. Decentralization of growth, the hallmark of competitive capitalism, was much less in China than Taiwan but it was still very significant. The estimates of TFP (total factor productivity) growth over the period 1979–1997 showed significant gains as follows:

China	1979	1997
Hong Kong	1.022	1.016
Guangdong	0.999	1.060
Fujian	1.014	1.053
Taiwan	1.030	1.027

2.3 New Ideas on Innovation Models

We discuss in this section several new ideas on innovation models as follows:

1. Stochastic models
2. Innovation matrix and diffusion
3. Aspects of disequilibrium.

Stochastic Models

Market selection process and how the firms innovate determine the survival of firms in an industry and the growth or decline of an industry. The theory of stochastic selection of industries for growth provides an important framework for analysis. It emphasizes that stochastic forces are vital in this process and it takes several forms. First of all, the decisions to invest for capacity expansion involve uncertainty about future demand and the possibility of future entry and technology competition. When investments are irreversible, the possibility of large sunk costs arises and this involves significant risks when the future demand fluctuations are expected to be high.

Schumpeterian theory emphasizes the innovation process in the market selection process. Technological innovations produce both substitution-cum-diffusion and evolution and these effects are generally nonlinear over time, path dependent involving multiple equilibria. This innovation stream has been frequently viewed as a stochastic process evolving over time. The transition of plants from one technology to another may be viewed as a birth and death process, i.e., a Markov process where birth may involve new plants or technology entering the system and death implying the plants close down due to obsolescence or competition.

We consider in this section two basic sources of stochasticity in industry growth. One is the stochastic birth and death process model, where innovation output (or efficiency) follows a creative destruction process as in the Schumpeterian framework. Secondly, the process of knowledge diffusion and learning affects the inter-sectoral growth process in a stochastic manner.

The stochasticity of the birth and death process depends on two parameters: the birth rate λ and death rate μ. The former represents new technology or new innovations, while the latter indicates the destruction or obsolescence of the old. If λ exceeds μ, then the innovation grows for the industry, leading to productivity growth and consequent price decline. This expands the market and globalization occurs. Stochasticity has two other effects. One is that the competition increases in intensity in the technology race and the firms struggle for survival of the fittest. For major and drastic innovations, the successful innovator may capture a more dominant position, and the others remain on the competitive fringe. This framework is most suitable for the leader–follower model. Alternatively, the framework may lead to rivalrous innovation, where the Cournot–Nash framework is more suitable.

Stochasticity has another important effect. It involves what is sometimes called the "churning process effect." This is like the creative destruction process which occurs when new entrants to the industry challenge the incumbents often with new innovations and as a result the exit rate rises. It has been empirically found that the higher the heterogeneity of the industry measured by output variance, the higher the exit rate. This results in higher concentration of large firms in the industry.

The stochastic birth and death process model may be simply modeled in terms of innovation effort $u(t)$ (e.g., R&D investments) by an innovative firm, where profit $\pi(n, u)$ depends on the number of firms n and $u = u(t)$. Sengupta (2011) has discussed the implications of this type of model for industry evolution. In this type of model the transition probability $p_u(t)$ of $u(t)$ taking a value u at time t satisfies the Chapman–Kolmogorov equation:

$$\frac{\mathrm{d}p_u}{\mathrm{d}t} = \lambda_{u-1} p_{u-1}(t) + \mu_{u+1} p_{u+1}(t) - (\lambda_u + \mu_u) p_u(t),$$

where the birth and death rule parameters depend on the level of u. Birth rate parameter leads to positive growth (i.e., positive feedback) and the latter to decay (i.e., creative destruction) due to the introduction of new technology. If λ_u, μ_u are positive constants (i.e., linear birth and death process), then the mean value function $m(t) = \mathbb{E}u(t)$ and the variance $v(t) = \mathrm{Var}u(t)$ of the process can be written as

$$m(t) = u_0 e^{(\lambda-\mu)t}, \ u_0 = u(0)$$
$$v(t) = u_0 e^{(\lambda-\mu)t} \left[u_0 e^{(\lambda-\mu)t} - 1\right].$$

Note that as the mean level of innovation rises, its variance increases over time more than the mean. An interesting case arises when the birth rate parameter λ_u declines with increasing u (e.g., the R&D field in new innovation is saturated) but the death rate is proportional to u^2 (i.e., due to the churning effect), i.e.,

$$\lambda_u = u a_1 (1 - u), \ \mu = a_2 u^2.$$

Then the mean value function follows the trajectory

$$\frac{\mathrm{d}m(t)}{\mathrm{d}t} = (a_1 + a_2) \left[\frac{a_1}{a_1 + a_2} m(t) - m^2(t) - v(t)\right].$$

This shows that the variance function has a large negative impact on the rate of change of $m(t)$. This is what is predicted by the churning process effect.

An interesting interpretation of the birth and death process has been given by Agliardi (1998) where the firms have a choice problem: which technological standard to chose, when there are two substitutable standards (e.g., two softwares) in the field, denoted by zero and one. There are N firms in the industry, and let $n(t)$ have standard 1 and $(N - n)$ have standard 0. Let $y = n/N$ be the proportion of N firms with standard 1. It is assumed that there are benefits from compatibility, i.e., firms are able to exploit economies of scale in using a common supplier of a complementary good. Following Agliardi assume that $n(t)$ is a birth and death process with transition intensities $\lambda(y)$ and $\mu(y)$ for the transition $0 \to 1$ and $1 \to 0$, respectively. Then he has proved an important theorem that under very general conditions $z(t) = \lim_{N\to\infty} \mathbb{E}y(t)$ exists and satisfies the differential

$$\frac{\mathrm{d}z}{\mathrm{d}t} = (1 - z)\lambda(z) - z\mu(z),\ z(0) = y(0);$$

the fixed points of this equation (i.e., when $(\mathrm{d}z/\mathrm{d}t) = 0$) are the stationary solution $\bar{z}$ of

$$\bar{z} = \frac{\lambda(\bar{z})}{\lambda(\bar{z}) + \mu(\bar{z})}.$$

Under the assumption that $(\partial\mu(z)/\partial z) < 0 < (\partial\lambda(z)/\partial z)$ which involves growth, there exist two solutions: one asymptotically stable, other unstable. The stable solution indicates that the system converges to one of the two standards. However, volatility also remains.

We have so far discussed the implications of positive feedback for the industry evolution. But firms vary in industry evolution in terms of both size and distribution. If some firms have positive feedback and others negative due to diminishing returns, then the interaction between these two groups leads to a dominance of the positive feedback firms. Consider for instance two groups of firms with outputs y_i $(i = 1, 2)$ growing exponentially.

$$y_i(t) = y_{i0}\exp(\lambda_i t),\ i = 1, 2,$$

where λ may represent the difference of birth rate and death rate intensities. If $\lambda_1 > \lambda_2 > 0$ due to Schumpeterian innovation, then the growth rate of the mixture $y(t) = y_1(t) + y_2(t)$ follows the dynamic process:

$$\frac{\mathrm{d}\ln\dot{y}(t)}{\mathrm{d}t} = \left(\frac{\mathrm{d}\ln y(t)}{\mathrm{d}t}\right)\left[\lambda_1 - \frac{\mathrm{d}\ln y(t)}{\mathrm{d}t}\right];$$

clearly as $t \to \infty$, the total output tends to grow at λ_1 which is the relative growth rate of the more efficient group. The average gain in efficiency for the industry defined as $E(t) = (\mathrm{d}\ln\dot{y}(t)/\mathrm{d}t)\lambda_2$ follows then the time path

$$E(t) = s(\theta e^{-st} + 1)^{-1},$$

where $\theta = y_{20}/y_{10}$ and $s = \lambda_1 - \lambda_2 > 0$. Then $E(t) \to s$ at $t \to 0$. There the parameter s may be interpreted as the efficiency advantage of the higher efficiency type over the lower.

An important area of stochasticity arises due to the diffusion process of the innovation stream and the learning phenomena. Unlike the Marshallian diffusion process, Schumpeter's diffusion process assumes that output growth $(\dot{x}/x)$ of an innovation industry is proportional to the profitability of the new technology, subject to the constraint that unit cost depends on the scale of production of the new technology.

$$\frac{\dot{x}}{x} \geq \frac{\dot{x}_d}{x_d} \quad \text{and} \quad \frac{\dot{x}_d}{x_d} = b(D(p) - x).$$

Here b is a constant indicating adoption coefficient, dot is time derivative, and $D(p)$ the long-run demand curve for the new community introduced by innovation. If growth of capacity agrees with the growth in demand $\frac{\dot{x}_d}{x_d}$ and price $p = kc(x)$ is proportional to marginal cost, we obtain a balanced diffusion path as a logistic model:

$$\frac{\dot{x}}{x} = \alpha - \beta x,$$

where

$$\begin{aligned} \alpha &= b(d_0 - c_0 d_1 k) \\ \beta &= 1 - c_1 d_1 k \\ c(x) &= c_0 - c_1 x \\ D(p) &= d_0 - d_1 p. \end{aligned}$$

Clearly there exist here several sources of equilibrium output growth. First is the diffusion parameter. The higher the diffusion rate of new technology, the greater the output growth. Stochastic forces play an important role here. Second, if demand (x_d) rises over time and the innovator has a forward looking view of market growth, it stimulates capacity growth. The rational expectation model highlights the importance of forward looking view in stimulating industry growth. Thirdly, the learning curve effect enables innovating firms learn about the scale economies in demand and market growth and implies the adoption of new innovations which imply declining unit costs and prices. Such a decline stimulates the innovation and growth process further through cumulative causation. Finally, the marginal cost also tends to decline for new technology firms due to knowledge spillover across different firms and industries. Recently Thompson (1996) developed a Schumpeterian model of endogenous growth, which relates the market value of a firm to its current profits and to its R&D expenditures, where the firm's relative knowledge follows a stochastic differential equation. This differential equation shows that the mean and variance of the output process of the firm is negatively correlated. This implies that the stochasticity is an important source of instability in the innovation framework, when the innovative firm's output augments industry growth.

Modern innovations occur in many forms. Besides Schumpeter's analysis of six types of innovations two of the most important ones are rivalrous innovation and endogenous innovation. In each case two types of new technologies have dominated the modern industrial field, e.g., specific purpose technology (SPT) and general purpose technology (GPT). SPT are incremental processes rather than drastic changes. Software innovations belong to this category. GPT has significant scale effects. Recent improvement in iPhone and other communication technology has dramatically changed the world market for information technology.

In nonrivalrous innovation the firms cooperate to take advantage of economies of scale. The spillover effects of different firms' R&D are jointly utilized, and the overall impact may be welfare increasing. In rivalrous competition however the race for winning the innovation for new process or new product continues. Successful innovations arise as a result of a Poisson process with an intensity u. The probability that a firm innovates successfully during the period $\mathrm{d}t$ is $u\mathrm{d}t$. Since firms are assumed to have equal chances at the beginning of the time period $(t, t+\mathrm{d}t)$ the probability that any particular firm becomes a winner in the race is $(u/n)\mathrm{d}t$, where n is the number of firms. The expected monopoly surplus from winning R&D races in the time interval $(t, t+\mathrm{d}t)$ is $(su/n)\mathrm{d}t$, where s is the monopoly surplus due to quality improvement through innovation and the consequential price rises. Denoting variable costs by $v(u)$ and fixed costs of R&D by c_f, the firms' instantaneous expected profit may now be written as

$$\pi(n, u) = \frac{si}{n} - v(u) - c_f .$$

At each instant the innovating firm chooses the R&D intensity $u(n)$ maximizing the instantaneous profit function $\pi(n, u)$ yielding $u(n) = (s/n)l^{1/(\sigma-1)}$; we assume $v(u) = u^\sigma/\sigma$. The optimal profit function may then be written as

$$\pi(n) = a^{-1}\left(\frac{s}{n}\right)^a - c_f .$$

This shows that the optimal profit function is monotonically decreasing for increasing $n > 0$.

In rivalrous innovation the firms are likely to be either Cournot–Nash competitors or Stackelberg competitors (i.e., leader–follower). In the former case a firm's payoff from innovation depends on the number of other firms that innovate successfully. In the latter case the leading firm acts as a dominant player and the market dominance model may be more appropriate.

Innovation Matrix and Diffusion

Computable general equilibrium (CGE) models are widely applied in different countries to study the sectoral interdependence. Coupled with a standard input–output (IO) model this framework evaluates the quantitative impact of external shocks including changes in technology. DeBresson and Andersen (1996) and his associates have developed and used empirically an innovative-interaction matrix between sectors which are suppliers of innovative activity and the sectors which are users. The supplier industries make up the rows and the user industries make up the columns. Innovative activity is measured either by investment or by output.

As investment innovation in sector i is explained in terms of investment by destination, i.e.,

$$I_i = \sum_{j=1}^{n} u_{ij} I_j, \sum_j u_{ij} = 1, u_{ij} \geq 0,$$

the growth of output of sector i is then related to innovative investment I_i. When measured in terms of output, two types of hypothesis have been put forward about the innovative activity affecting different sectors of the economy. One is by Schumpeter who postulated that the innovations tend to concentrate in certain sectors due to agglomeration effects of economies of scale rather than evenly distributed over the entire economic space. DeBresson finds substantial evidence of such innovation clusters in the UK, Italy, and Greece. Recently the most successful NICs in Asia like China, South Korea, Taiwan, and Japan have displayed this concentration most significantly. A second trend is the close interdependence between the innovative activity and the sectoral linkages. DeBresson estimated a linear regression equation for Italy over the period 1980–1984 with innovative output (I) as the dependent variable and the following three independent variables: economic linkages both forward and backward (L), and index (T) of linkages with the world technology and the R&D expenditure (R):

$$I = -136.9 + 8.92L + 29.6T + 0.02R$$

$$\bar{R}^2 = 0.69.$$

Clearly this shows that the impact of foreign technical know-how and its diffusion is very important. For the successful NICs in Asia like South Korea, Japan, Taiwan, and Singapore this type of international diffusion of innovative knowledge has played a most significant catalytic role in their rapid growth episodes.

Aspects of Disequilibrium

Based on the neoclassical optimization model the Solow model analyzed long-run economic growth of per capita output through a steady-state model of equilibrium. Technology and innovations were exogenous in this formulation and convergence to the steady-state equilibrium was guaranteed by the assumption of diminishing returns to capital. Modern theory of endogenous growth assumes endogenous technology, where knowledge capital and innovation have increasing returns to scale. The spillover of information technology and innovation linkages across countries have been emphasized by Romer, Lucas, and other growth as critical in explaining the rapid growth episodes in the successful NICs in Asia. The case of Taiwan provides a notable example. It shows the intensity of knowledge diffusion and software development to a significant degree. Also Taiwan's record of performance in the IT sector is most impressive. In terms of the average number of annual US patents per million people, the top rankings in the world in 2004 are 1 for the USA, 2 for Japan, and 3 for Taiwan. The numbers of patents are 301.4 (USA), 273.4 (Japan), and 241.4 (Taiwan).

Unlike the steady-state formulation of growth in the Solow model, the Iwai model of diffusion emphasizes the non-steady-state and disequilibrium properties due to innovation. In Schumpeterian dynamics innovations not only tend to be concentrated in certain sectors, e.g., the IT sector in modern times, but they tend to disrupt the system and create disequilibrium. Entrepreneurs who carry out the innovation move the whole industrial system away from the neighborhood of equilibrium to the upside. Economic evolution is characterized by upward-moving neighborhoods of equilibrium that are separated from one another by two distinct phases. In the first the system draws away from equilibrium under the impulse of innovations and during the second it moves to another equilibrium.

The analysis of non-steady-state growth has striking similarity with the genetic theory of evolution. This theory emphasizes the fitness principle underlying evolution. Evolutionary economic theory has used the replicator dynamics principle of genetic evolution model to explain the variety of pattern of industry growth. The central hypothesis of this model is that the frequency of a species (e.g., technologies or firms) grows differentially according to whether it is below or above the average fitness. If genetic fitness is replaced by economic efficiency or core competence, the replicator dynamics in firm growth can explain the industry evolution and growth. Metcaffe (1994) and Mazzucato (2000) have used this type of replicator dynamics to explain the non-steady-state pattern of industry evolution following a Schumpeterian innovation framework. This innovation process may be viewed as affecting the entry and exit behavior of a dynamic market and its growth. Thus if $N(t)$ be adoption of new innovations (e.g., new combinations, new products, or new processes). One may then formulate innovation as a diffusion process:

$$\frac{\mathrm{d}N(t)}{\mathrm{d}t} = \left(p + \frac{qN(t)}{m}\right)(m - N(T)).$$

Here m is the ceiling of $N(t)$, p the coefficient of innovation, and q is the coefficient of imitation. Assuming $F(t) = N(t)/m$ the fraction of potential adopters who adopt the technology or the innovation at time t, one type of diffusion model is of the form

$$\frac{\mathrm{d}F(t)}{\mathrm{d}t} = (p + qF(t))\,(1 - F(T));$$

if $p = 0$ then we consider the imitation effect alone, where firms tend to imitate the invention process of others as in the leader–follower model. If $\theta = 0$ and $F_0 = 0$, then $F(t)$ and hence $N(t)$ follow a logistic path that has been empirically observed for many technological innovations. Note that the ceiling m itself may increase in the long run and the growth process may be explosive.

Schumpeter's innovation approach to evolutionary economics which emphasizes non-steady-state and disequilibrating aspects of industry growth contains five elements as follows:

1. The rate of growth of innovations depends on two factors: the gap of the advanced technology from the existing one.
2. Unit cost declines due to new investment in innovation. This results in excess quasi-monopoly profits for the successful innovators who adopt the advanced technology.
3. Profits from innovative investment lead to further innovations, which generate diffusion of new knowledge. This diffusion and spillover linkages provide incentives for other firms and industries to innovate further.
4. Birth rate of new innovations through R&D investments provides the positive side of industry evolution, where the death rate provides the negative side. The contagion effect of birth rates influences other firms to invest in more R&D.
5. Long-term profit maximization and faster rates of growth provide the basic incentives for firms to innovate.

The relations above can be used to develop a dynamic model of industry growth with non-steady-state characteristics, where the trajectories would indicate the paths of endogenous evolution. Scale economics and profit incentives are the two key forces in this growth paradigm. Unbounded growth and oscillations may occur in this path-dependent growth trajectories.

Chapter 3
Schumpeterian Innovation

Schumpeterian innovation theory uses three basic premises for industry growth: creative accumulation, creative destruction, and rejection of competitive market equilibria. These aspects all emphasize evolutionary growth and endogenous innovation. Market dynamics, profit expectations, and long-run growth provide the key ingredients of this innovation theory. We attempt a critical review of this theory in this chapter as follows:

1. Creative accumulation
2. Creative destruction
3. Adjustment mechanisms
4. Evolutionary fitness
5. Dynamic flexibility

3.1 Creative Accumulation

Schumpeterian concept of creative accumulation has several dynamic features of which four are most important: "new combinations," "entrepreneurship," "dynamic flexibility," and "the role of the banker" in facilitating finance in investment.

The implementation of new combinations in a discontinuous manner over time may take different forms. Developing a new source of supply, establishment of new firms, or change in the production function are some examples. The new combinations use idle capacity and reallocate resources to increase economic efficiency. Heertje (1988) emphasized that the new combinations of Schumpeter involved discontinuous bursts and diffusion among interdependent firms. In most cases the new combinations do not drive out the old firms but compete with them. New combinations view innovation as the setting up of a new production frontier.

The individuals who carry out this innovation are called the entrepreneurs. They carry out an innovation with an expectation for profit. The profitability of an innovative project depends according to Schumpeter on the conservatism of the

J. Sengupta, *Theory of Innovation: A New Paradigm of Growth*,
DOI 10.1007/978-3-319-02183-6_3,

old firms and the speed at which both imitative and/or innovative firms emerge. One has to distinguish between two concepts of entrepreneurs: the Walrasian and the Schumpeterian. Andersen (2011) discussed this point in some detail. His analysis stressed the point that the Walrasian equilibrium concept has the apparently paradoxical characteristic, i.e., his W-entrepreneurs who are driven by the profit motive obtain neither gain nor loss when the economic system is brought into equilibrium through a process of trial and error (tâtonnement). Schumpeterian changes the given production routines in order to earn higher expected profits which are positive. Schumpeter's distinction between innovation and noninnovative agents provided him with the theory of change of the data and the given production system. The S-entrepreneur disturbs the Walrasian equilibrium by buying and using resources to change one or more of the parameters of the economic system. The S-entrepreneurs provide the key agents of the Schumpeterian evolutionary model. This model had five basic elements:

1. Process types: The economic process has two basic elements: the static and the dynamic. The dynamics involve both adaptive and new innovations.
2. Evolution: The dynamics involve both internal and external changes in the economy. Evolution focuses on selection of fitness among firms as in genetic evolution.
3. Disturbance: Economic development due to innovation is basically a process of disturbance of the static general equilibrium of the economy.
4. A new equilibrium: The disturbance caused by innovation initiates a new process which normally generates a new equilibrium.
5. Reorganization: The shift to a new equilibrium generally involves a reallocation of resources and a reorganization of the cost price system.

The Schumpeterian creative accumulation involves five basic types of innovations discussed in some detail by Andersen:

1. Product innovation
   A new type of product or service is added to the existing system requiring a new production routine and also a change in consumption network.
2. Process innovation
   A new technology for an existing product, requiring changes in input and output qualities.
3. Organizational innovation
   Economies of scope and scale involving changes in business organization and strategies for new market structures.
4. Market innovation
   Changes in market structure involving globalization of trade, e.g., variants of iPhone introduced by Apple.
5. Input innovation
   A new raw material or a new intermediate good is introduced into the economic system, e.g., software development. This may frequently involve the opening up of new sources of supply.

These different types of innovation serve to increase the complexity of the evolutionary model of Schumpeter's growth theory. They also serve to generate innovation-induced disequilibria. From the viewpoint of agents following the routinized equilibrium, the innovative entrepreneur challenges the functioning of old routines. New firms emerge based on new routines and the renewed equilibrium forms the basis for another phase of disequilibrating innovative economic activity. Thus the long-term economic evolution consists of a series of routinized equilibria and the innovative strategies of new firms that disturb these equilibria. Thus the Schumpeterian scheme provides a model of a wave form economic evolution.

A dynamic aspect of Schumpeter's theory of innovation is the concept of "induced innovation" which denotes those additional improvements which present themselves in the process of copying the first innovators in a business field and consequent adaptations by the existing firms. Then adaptations also generate both potential and actual expansion of total industry output.

3.2 Creative Destruction

Schumpeter discusses two grand paradigms of economic growth: one is the equilibrium economics of the Walrasian system and the other is the evolutionary economics. The second paradigm contains the essence of real economic development, i.e., how does an economy make the transition from one level to another.

Two basic dynamic forces operate here. One is the concept of creative destruction used frequently by Schumpeter and the other the concept of dynamic selection often used in the fitness model of genetic evolution. We discuss now these two aspects of Schumpeterian model of evolutionary growth.

The reactions and response of Schumpeterian entrepreneurs (S-entrepreneurs) are very critical to Schumpeter's theory. If the reactions are slow, then their selling prices remain above costs for a relatively long period and the average prices may increase due to additional demand created by a group of innovative projects. However, when this type of demand disappears, the average prices will fall below the costs of at least those following old routines and their economic responses are either adaptation or bankruptcy. This is what Schumpeter called "the perennial gale of creative destruction." Another type of creative destruction occurs through economic competence. Firms innovate to augment their competence and efficiency. If they succeed, then the old firms following old or existing routines cannot withstand the pressure and as a result they are forced to exit. Schumpeter's notion of technology is much broader than is represented in many econometric applications and the diffusion process is accordingly broad. The waves of innovation which serve as generation of long-term economic growth are innovations of wide applicability. These all dramatically widen the internal opportunity set, allowing for a great range of new commercial combinations to develop. This also serves to accelerate the destruction process more widely. Finally, we have to note that Schumpeter emphasized the point that the complicated process of returning to equilibrium from a highly disequilibrated state puts the brake on further innovation. This helps the process of reaching a new equilibrium.

Schumpeter's theory of economic evolution has been discussed in some detail by Andersen (2011) in terms of three models labeled as Mark 0, Mark I, and Mark II. Phlips (1995) and later Freeman (1982) emphasized Schumpeter's distinction between innovation through new firms in his book *The Theory of Economic Development* (1934) and innovation as part of oligopolistic competition in *Capitalism, Socialism and Democracy* (1942).

We follow Andersen's analysis of the three models.

The Mark 0 model of pure adaption. If undisturbed this adaptive economic process ultimately reaches a stationary state. This is very similar to the neoclassical paradigm. It is this state that serves as the entry point of Mark I's innovative entrepreneurs. Adaptation takes place under the condition of competitive struggle between firms involving selection and learning in which some firms go bankrupt while the others succeed in introducing the new innovation and change.

Mark I adaption occurs through innovation involving a creative response by the S-entrepreneur who calculates the expected profitability of the innovative project and determines the optimal timing. Economic evolution takes place in two phases. After an upswing is caused by high innovative activity follows a downswing characterized by enforced adaptation.

Mark II model is one of oligopolistic adaptations. This model depicts the oligopolistic struggle for competitive markets. It does so by removing Mark I's assumption of the conservativeness of incumbent firms. This model is especially suitable for analyzing oligopolistic competition and in fighting between innovators and imitators in Cournot-type or Bertrand-type oligopolistic industries. Large firms play a dominant role here.

Schumpeter's three models of economic evolution all assume that the socioeconomic coevolution also influences economic evolution parametrically. The forces of creative destruction are basically complementary to the process of creative accumulation.

3.3 Adjustment Mechanisms

The Schumpeterian models of economic evolution are characterized by routine, innovation, and adaptation. Schumpeterian innovation introduces new routines in place of the old. The new routine spreads and diffusion starts if the competitive and/or oligopolistic process favors the adopters of new routines and if those incumbents bound by the old routine are competed out of the system. The mechanism of adaptation is thus a mechanism of selection between routine. Thus inertia, innovation, and selection operate in interrelated dynamics to define an economic evolution. Andersen discussed the total evolutionary change and the adjustment mechanism as the change in the weighted average routine at the industry level comprising four effects as follows:

$$\text{Total evolution} = \text{Entry effect} + \text{Selection effect} + \text{Intra-firm change effect} + \text{Exit effect}$$

The entry effect initiates the new innovation. It starts the competitive struggle with the incumbents. The selection process emphasizes the fitness principle of genetic evolution and the intra-firm change effect generates substantial reallocation of old and new resources. The exit effect secures the balance through new industry equilibrium.

There is a fundamental difference between the two types of entrepreneurs, the Walrasian (W) and the Schumpeterian (S), in the adjustment mechanisms. To explore profits these two entrepreneurs respond differently. For W-entrepreneurs positive or negative profits occur as long as the economic system is in disequilibrium. But according to Walrasian adjustment this disequilibrium is removed by perfect competition among W-entrepreneurs. When this competition has brought the system into equilibrium through a process of trial and error (tâtonnement) the W-entrepreneurs make neither gain nor loss. The S-entrepreneur explores the profit motive by innovative investment. He disturbs the equilibrium by buying and using resources to change one or more of the parameters of the economic system.

Schumpeter considered six sets of innovation of which the introduction of a new method of production or technology and the opening of a new market through the introduction of new products are the most important. The dynamic role of R&D investment in the strategy followed by a successful innovator has been especially emphasized by Schumpeter. In this approach the successful innovator produces the newly invented good and protected by patent rights he may succeed in driving out the previous incumbents by undercutting their price. Thus the successful innovator enjoys monopoly rents until driven out by the next innovation. Two aspects of the innovation process play a key role in Schumpeterian dynamics. One is the existence of learning by doing by which the first movers among the innovating firms race down the learning curve to gain a cost advantage. If the cost advantage is large, its possessor may select to set prices low enough so as to deter new entry of other competitors. Thus the innovating firms have an incentive to lead by introducing new products or new designs which involve learning by doing. Also, the aggressive pricing of new products may stimulate the demand both intensively and extensively. Schumpeter's emphasis focuses on diffusion effects as an evolutionary process. If $x(t)$ is the supply of a new commodity at time t and x_{d} is demand, then this diffusion model assumes that output growth is proportional to the profitability of new technology, subject to the constraint that unit cost depends on the scale of production of the new technology:

$$\dot{x}/x \geq \dot{x}_{\mathrm{d}}/x_{\mathrm{d}}$$

and

$$\dot{x}_{\mathrm{d}}/x_{\mathrm{d}} = b(D(p) - x)$$

Here b is the constant adoption coefficient, dot is the time derivative, and $D(p)$ is the long-run demand curve for the new commodity, where $D(p)$ rises when unit price p falls due to scale effects. If the growth of capacity is in equilibrium with demand growth and price is proportional to marginal cost $p = kc(x)$, we obtain a balanced diffusion path as a logistic process:

$$\dot{x}/x = (\alpha - \beta x)$$

where

$$\alpha = b(d_0 - c_0 d_1 k), \beta = 1 - c_1 d_1 k$$
$$c(x) = c_0 - c_1 x, D(p) = d_0 - d_1 p$$

Here there exist several sources of equilibrium output growth. First is the diffusion parameter b. The higher the diffusion rate of new technology, the greater the output growth. Second, if demand expands due to price cost decline and the innovator has a forward-looking view of market growth, it stimulates growth in capacity further. Third, the marginal cost tends to decline due to knowledge spillover across different firms and industries on an international scale. Finally, stochastic variations in the diffusion and technology adoption parameters are very basic to Schumpeterian dynamics, which views overall industry growth as arising from the flow of industrial innovations and knowledge. The Iwai model discussed in chapter two extended the Schumpeterian dynamics of interaction between technological innovation and diffusion. The Iwai model showed that the diffusion parameter b can be approximately expressed as

$$b \sim \beta_s \rho w / k_o$$

where β_s is the constant propensity to save out of profits, w is the share of wages in national income, k_o is the capital–output ratio, and ρ is the constant speed of adjustment. Thus the speed of diffusion depends on the firm's responsiveness to the profitability of alternative technologies.

Another evolutionary aspect of the logistic growth path is that the change in upper asymptote of the growth path due to the long-run impact of large-scale innovation may be viewed as an evolutionary shift of the productivity frontier.

3.4 Evolutionary Fitness

Schumpeterian theory of evolution emphasized in his long-run economic wave models used genetic evolution theory for comparing the outcomes or values of two or more technological routines. Schumpeter had been fascinated by Galton's statistical study and Darwinian natural selection, but Schumpeter failed to apply Ronald Fischer's approach to genetic evolution through fitness.

In Schumpeterian theory the selection process between routines promotes and demotes firms based on their characteristic values or size. This promotion and demotion is analyzed in terms of the so-called fitness of firms. The effect of selection between routines is due to the covariance (cov) between fitness (f) and values (z), i.e.,

$$\text{Effect of selection between routines} = \text{cov}(f, z) = \beta(f, z)\text{var } z$$

Here $\beta(f, z)$ denotes the efficiency of selection and var z measures the ease with which selection can take place in the preselection population. In Schumpeter's theory the selection efficiency is important, since it measures the degree to which variance can be exploited by innovation to produce economic evolution.

This line of thought has been recently extended in modern theory of endogenous economic evolution. Schumpeter defined evolution not simply as change but as a process of *endogenous* change and recognized that an application of Darwinian concepts to economic evolution was potentially useful. Kelm (1997) has discussed in some detail the role of Darwinism in Schumpeter's theory of economic evolution. We discuss here in brief Kelm's ideas.

A Darwinian theory explains the change in the distribution of characteristics in a population by the interaction of three mechanisms operating at the individual level as:

1. A mechanism of information storage by which some relatively stable characteristics are preserved over time
2. A mechanism of endogenous change by which new variations are constantly generated
3. A mechanism of selective retention by which the frequency of some variations relative to others are increased

It is important to discuss these mechanisms in brief in relation to economic evolution.

The idea that in economic evolution routines and habits play the same role as genes or genetic inheritance plays in biological evolution is familiar to evolutionary economists. As Nelson and Winter (1982) pointed out these routines and behavioral foundations of economic evolution are inconsistent with the notion of rational optimization in the Walrasian equilibrium model. The Walrasian equilibrium built through the assumption of rational optimization was considered highly artificial by Schumpeter.

The sources of endogenous change that drive biological evolution are mutation and recombination. In Schumpeter's view innovation is the fundamental mechanism of endogenous change driving economic evolution. This innovation may come from the new consumer goods, new methods of production, or the new forms of industrial organization that the capitalist enterprise creates. Here Schumpeter anticipated the concept of "genuine uncertainty" resulting from a course of action associated with an innovation. He stated unequivocally that despite the higher degree of "conscious rationality" involved in innovation than in routine action, the choice of new methods is not simply an element in the concept of rational economic action but a distinct process requiring special explanation. This special explanation of the innovation mechanism consists in the phenomenon of entrepreneurship in Schumpeter's view. This entrepreneurship involves "leadership" which is required to overcome the resistances by industry groups potentially threatened by the innovation. The profit motive and its role in innovation dynamics are central to Schumpeter's view of economic evolution. Thus he not only emphasized the Walrasian result absent in a stationary process but also the point that successful innovations usually translate into profits in an evolutionary process.

Selective retention is the third important characteristic of human agency or entrepreneurship. As soon as an entrepreneur has successfully introduced an innovation, the possibility of its imitation by others arises. Other entrepreneurs follow and sometimes in increasing numbers, and the path of innovation becomes progressively smoothed by accumulating experience and learning; the recent history of iPhones is a familiar example. Schumpeter's discussion of the learning process following innovation implicitly identifies the evolutionary force of what is called "biased transmission" by Boyd and Richardson (1985). Transmission is biased if the process of transmission itself is a process not of random but of guided variation. Schumpeter identified the biased transmission of innovations as a central force of selective retention in economic evolution. For this reason the aspect of natural selection that seems most useful to apply to economic evolution is *differential survival*. Some important necessary conditions for natural selection are fulfilled in market economics as follows:

1. Due to the mechanism of information storage in economic evolution, different firms are characterized by different sets of relatively stable routines, and these sustained variations are usually associated with different chances of survival in a market environment.
2. The idea of differential survival is at the heart of Schumpeter's notion of competition, which is different from the Walrasian concept. In a modern capitalistic framework it is not the Walrasian price competition that counts. What counts is the competition from the new commodity, the new technology, the new source of supply, and the new type of organization. Differential survival is the end result of this type of competition.

Innovations are as a rule embodied in new firms which carry on new ideas for the market. This is a cumulative process. Learning begets experience and experience begets new ventures in knowledge capital. The similarity of Schumpeter's notion of competition with Darwin's concept of natural selection stands out clearly when we analyze the evolutionary aspect. In genetic evolution the importance of historical contingency has always been recognized. This historical contingency is for example due to the influence of chance events on natural selection and to the phenomenon of path dependence. Schumpeter ascribed the same properties to the process of economic evolution.

Schumpeter discussed in detail how the complex dynamics of firms in capitalist economies result from the interaction of various Darwinian forces implicit in the evolutionary theory. He analyzed how the economic system responds to the intrusion of an innovation, e.g., for some of the old firms, new opportunities for expansion open up. This indicates that the innovations have positive externalities. But for other firms the emergence of new methods or products means economic death. Creative destruction occurs along with creative accumulation. Expected profits constitute the adoptive standard for the evolutionary forces dependent on entrepreneurial innovation.

3.5 Dynamic Flexibility

Economic progress results mostly from innovations relating to the production process, as distinct from product innovations. These innovations serve directly to improve productive efficiency. As Schumpeter noted, capitalism is essentially a process of endogenous economic change. Without this change the capitalist society cannot exist. Klein (1988) has discussed in some detail the concept of dynamic flexibility underlying Schumpeterian productive efficiency. To quote Schumpeter (1942),

> A system-any system, economic or other-that at every given point in time fully utilizes its possibilities to the best advantage may yet in the longer run be inferior to a system that does so at no given point of time, because the latter's failure to do so may be a condition for the level or speed of long-run performance.

Klein relates this statement to dynamic flexibility in the context of a positive sum game. Dynamic flexibility has to be distinguished from complete specialization. Specialization implies nearly perfect adaptation to an existing environment but dynamic flexibility in the ability to make speedy adaptations in the face of new situations. Consider the example of Sony. To take best advantage of existing possibilities it would need to build plants in a manner so as to minimize production costs at a given point of time, but Sony does not do that. Why? Because it will tend to reduce further innovations. Full automation of all plants will restrict the scope of further learning to continue new innovations.

What dynamic flexibility buys in terms of productivity gains can be more rationally expressed in terms of positive sum games. This has been discussed in some detail by Klein (1988). He introduced the concept of "cooperative dynamic competition" to describe such a positive sum game. Suppose that firms in an industry operate on the expectation that as a result of their cooperative combined efforts to improve cost efficiency through pooling risk taking, a more than proportional sales increase will take place. In such an industry there will be winners and losers in every round of the game, but it is a positive sum game. This type of cooperative dynamic competition may help internalize a large portion of benefits from innovative R&D investments which spillover. By contrast the zero-sum games where gains and losses cancel each other, are not likely to feature a significant degree of risk taking. As a result the volume of innovative investment would be much lower.

This framework is closely related to the waves of innovation commonly referred to as generators of long-term economic growth. Eliasson (1988) has discussed in some detail various examples of technologies with universal applications, which create giant expansion of the opportunity set and also expose firms in increased competition. This also generates mergers across national boundaries. The standard examples include steam engine, electricity, personal computers, and software technology. These all dramatically widened the international opportunity set allowing for a great range of new commercial combinations to develop. But the final economic outcome of this expansion of opportunity set depends on the local competence and dynamic flexibility of new innovative firms.

3.6 Innovation and Market Competition

There are many ways by which the Schumpeterian innovations change the market structure. New markets, international expansion, and rivalry from old competitors create new market dynamics. D'Aveni (1994) has characterized this dynamic as hypercompetition with rivalrous competition. Two dynamic strategies have been stressed by his model. One is the critical advantage of the first mover or first innovator who enters the market with new innovation and sustains competition successfully. The second is the dynamic flexibility which the successful innovator must pursue over time. Thompson (1996) has developed an endogenous model of industry growth based on the Schumpeterian notion of expanding technological opportunity set due to innovative firms. Schumpeter discussed two forms of endogenous growth under capitalism. One is the *competitive capitalism*, in which innovations emerge within entirely new firms, driving existing firms out of business. A second form of competition is called *trustified capitalism*, where firms are very persistent in the face of innovations by others. Here innovation goes on with the big units now existing largely independently of others. Under trustified capitalism, firms' market share may respond to innovations by themselves and by others but market participants are likely to be relatively stable over time.

We discuss now in some detail the model of trustified capitalism developed by Thompson, who applied this model to firm-level data for the period 1973–1991. This model predicts a simple relationship between stock market valuations, R&D expenditures, and profits. The relationship between stock market values and current profits is assumed in this model to depend on the market's expectations of the long-run rate of growth of knowledge or equivalently the rate of creative destruction in the economy. We discuss below in some detail this dynamic equilibrium model due to Thompson.

The economy is made up of a continuum of firms indexed by $i \in [0, 1]$ each producing a differentiated good. Each firm maximizes profits at each time point t conditional on its current level of knowledge relative to the rest of the economy. Firms invest in an attempt to increase their relative level of knowledge defined as

$$\alpha(i,t) = q(i,t)/Q(t) \tag{3.1}$$

where $Q(t)$ is the economy-wide knowledge index. Innovations which increase a firm's level of knowledge occur in a stepwise fashion at random intervals; the expected length of which is decreasing in R&D expenditure. As the firm innovates, nominal knowledge $q(i,t)$ are assumed to be independent homogeneous Poisson jump processes with intensity parameters $\lambda(i,t)$ and jump magnitude $\gamma(t)Q(t)$.

On differentiating (3.1) we obtain a stochastic differential equation which governs the evolution of the firm's relative knowledge level

$$d\alpha(t) = -\alpha(t)\beta(t)dt + dq^*(t) \tag{3.2}$$

where $\beta(t) = \dot{Q}(t)/Q(t)$ is the economy-wide growth in knowledge and $q^*(t)$ is the Poisson jump variable.

Let $\pi(\alpha(t))$ be the rate of profit gross of R&D expenditure of the firm and it is assumed that competition between firms is Schumpeterian in the sense that the primary determinant of profit is the relative level of knowledge. Now the firm's decision problem under trustified capitalism is to maximize its discounted profits by an approximate choice of R&D effort $R(t)$:

$$\max_{R(t)} E_t \int_t^{\infty} e^{-\rho t} \left[\pi(\alpha(\tau)) - w(\tau)R(\tau)\right] d\tau$$

subject to (3.2). Here E_t denotes the expectation conditional on the observed value of $\alpha(t)$ and $w(\tau)$ in the unit cost of R&D effort. Explicit solution of this problem is very complicated and dynamic programming algorithms are needed. However in the special case when the profit function is linear in the level of knowledge and R&D production function is isoelastic,

$$\lambda(t) = R(t)^{b(t)}, 0 < b < 1$$

$$\pi(\alpha(t)) = A(t) + B(t)\alpha(t)$$

the profit maximizing intensity of R&D may be explicitly computed as

$$R(t) = \left[\frac{b(t)B(t)\gamma(t)}{w(t)(\rho + \beta(t))}\right]^{1/(1-b(t))} \tag{3.3}$$

Note that $R(t)$ is independent of the level of knowledge. On differentiation of (3.3) it follows that the optimal intensity of R&D varies directly with the magnitude of the Poisson jumps $\gamma(t)$ and with the elasticity of the R&D production function. Also, the R&D level varies inversely with the firm's unit R&D costs and the firm's cost of capital. Finally, an increase in the economy-wide rate of technology growth induces a reduction in the R&D intensity of the firm. The evolution of R&D intensity over time depends on the evolution of the variables $b(t)$, $B(t)$, and $\gamma(t)$.

On using the value function $V(t)$ from the Hamiltonian–Jacobi–Bellman equation for the firm's profit maximization problem one could write for our special case

$$V(t) = a_0 + a_1\pi(t) + a_2 w(t)R(t) + \epsilon(t) \tag{3.4}$$

where $\epsilon(t)$ is an additive error term and $V(t)$ is the value function given by

$$\rho V(t) = \max_{R(t)} \{\pi(\alpha(t)) - w(t)R(t) - \alpha(t)\beta(t)V'(\alpha(t)) \\ +\lambda(R(t))\,[V(\alpha(t) + \gamma(t)) - V(\alpha(t))]\}$$

The value function may also be written as

$$V(t) = \frac{\beta(t)A(t)}{\rho(\rho + \beta(t))} + \frac{1}{\rho + \beta(t)}\pi(t) + \frac{(1 - b(t))}{b(t)\rho}w(t)R(t) + \epsilon(t)$$

The Eq. (3.3) provides a convenient way to statistically estimate the relation between R&D and the value of the firm, because the relationship is linear between R&D expenditure $w(t)R(t)$ and the value of the firm.

Thompson estimated (3.3) using Standard and Poor's cross-section Compustat database for 19 firms for the period 1973–1991. The number of observations varies between 730 firms in 1973 and 2,412 in 1991. The firms which are all American-owned publicly traded corporations are divided into 13 major industries according to Standard and Poor's primary activity classification and the elasticity of R&D production function was allowed to vary across industry groups. In contrast the jump magnitudes of innovations are constrained to be equal across industries. This constraint allowed one to distinguish between economy-wide changes and changes in the industry level innovations due to R&D expenditure. Market value was measured as the sum of market capitalized value and outstanding debt and profits were measured by annual income gross of R&D expenditure.

The economy-wide parameters, the expected rate of growth in knowledge $\beta(t)$, the values of $A(t)$ and $B(t)$, the value of an innovation, and the Poisson jump parameter $\gamma(t)$, are related to the coefficients of Eq. (3.3) as

$$\beta(t) = (a_1(t))^{-1} - \rho$$

$$A(t) = (a_0 t\rho)(1 - a_1(t)\rho)^{-1}$$

$$B(t) = N(t)\beta(t)\left(\sum_{j=1}^{13}\sum_{i=1}^{N_j(t)} R_{ij}(t)^{b_j(t)}\right)^{-1}$$

where $i = 1, 2, \ldots, N_j(t)$ and $j = 1, 2, \ldots, 13$ and $\pi = A(t) + B(t)\alpha(i, t)$. The cross-section estimates for selected years are as follows:

Several comments are in order. First, the estimates of $\gamma(t)$ provide an index for the combined behavior of appropriability and advances in knowledge. While $\gamma(t)$ shows a sustained decline during the 1980s, it had experienced significant growth during the 1970s. Second, the average value of innovation is around 2.30 million. The value of a typical innovation shows a steady decline during the sample period from a peak of $4.7 million in 1973 to 0.91 million by 1991. This may be due in part to a change in market structure as the creative destruction emphasized by

Table 3.1 Selected parameter estimates of model (3.3)

	1973	1980	1987	1991	Mean
Growth (%)	3.53	9.02	4.79	3.88	4.497
A ($M)	7.28	2.08	1.15	0.59	2.30
$B(t)$ ($M)	150.1	148.9	88.3	56.0	110.7
$\gamma(t)$ (1,000)	1.41	1.83	0.56	0.79	0.99
Innovation (value) ($M)	4.70	2.72	0.85	0.91	1.96
Technological opportunity index (5 out of 13 industries)					
Food and Textiles	0.39	0.5	–	–	–
Chemicals	0.50	0.59	0.86	–	0.65
Industrial equipment	0.87	–	1.31	1.09	0.76
Electrical equipment	0.40	0.58	0.94	0.92	0.64
Transport equipment	1.35	0.89	1.17	0.80	1.11

Table 3.2 Technological opportunity rankings

	Ranking by opportunity	OECD technology level	Rankings by growth rates
Industrial equipment	1	H	1
Instrument manufacturing	2	H	3
Transport equipment	3	H,M	4
Business services	4	–	6
Chemicals	5	M	7
Metallurgical industries	6	L	8
Glam manufacturing	7	–	5
Communications	8	–	10
Food and Textiles	9	L	9
Electrical equipment	10	H,M	2
Mining	11	–	11
Wholesale business	12	–	12

Schumpeter. Third, the technological opportunity index which measures the ease of introducing innovations is seen to vary widely between industries. It is interesting to observe that the behavior of each industry's technological opportunity index over time is largely independent of what is happening in other industries. The index is as high as 1.28 in the industrial equipment group which includes computers (Table 3.1).

Table 3.2 compares industry rank by technological opportunity with the OECD estimates which classify industries according to whether they are high (H), medium (M), or low (L) technology industries. Thompson's technological opportunity rankings compare favorably with the OECD assessments. One has to note however that the services sector is ranked relatively highly by technological opportunity suggesting higher rate of innovation in that sector.

Thompson also estimated the expected rates of knowledge growth for the 13 industry groups. Over the period 1973–1991 the average expected annual rate of growth ranged from about 0.8 % for the retail sector to 8.7 % in the industrial equipment sector. The rates of growth are affected by three main factors: changes in $\gamma(t)$, change in industries' technological opportunities, and changes in R&D intensity. Thus while the electrical equipment manufacturing sector was ranked low by technological opportunity, it compensated for this with a high intensity of R&D, thereby generating an average expected growth rate of 7.9 %.

Two comments are in order by way of concluding remarks. One is that Thompson's model of Schumpeter's trustified firm used an operational framework for identifying knowledge growth as the critical factor of Schumpeterian innovations. This growth comprises both productivity and quality gains. Secondly, the appropriability of the benefits of innovations is found to be very important in Thompson's formulation of the Schumpeterian model. The spillover effects emphasized in the current theories of innovation suggest that the total innovation and R&D at the industrial level can be considerably increased if a degree of cooperation or state subsidy could be applied. This has been discussed by Spence (1984) and others.

3.7 Sharing Knowledge Capital

In modern high-tech industries like computers, telecommunications, and pharmaceuticals R&D investment for improving technology through innovation has played a dynamic role. This type of investment has significant economies of scale both internal and external. Internally it means that a firm's unit cost declines as the size of this investment increases. By pooling such investments the industry can lower its average cost of production. This leads to price declines and hence an expansion of demand. External economies refer to the spillover effects on other firms so that the total industry outcome is lower unit costs and hence prices.

In today's world business the expansion of total industry output may be generated by rapid growth in world demand due to demand-side economies of scale. For example, the users of Microsoft value its operating systems because they are widely used in the industry. Unlike the supply-side economies of scale, demand-side economies of scale do not get exhausted when the market expands more and more. In dynamic competition this demand-side economies of scale may generate a substantial shift of the industry demand curve upwards. Three aspects of this demand expansion have played important roles in recent years. One is the emergence of globalization of trade. Adam Smith strongly emphasized the role of competitive international trade and its expansion as the prime mover of industrial growth. The second aspect is the growth of knowledge capital in the industry through computers and other telecommunication tools, which has increased the efficiency inputs of an average firm. Knowledge capital and innovations have proved to be complementary to labor and other physical inputs to the firm and prevented diminishing returns to operate in the production process. Finally, an expansion of the knowledge capital and hence the industry output generally lead to a more efficient labor force with consequent reductions in unit costs for the firm and the industry as a whole. This helps the dynamic process of overall industry and economy growth.

Innovation and the evolution of industries have witnessed major progress in recent years. Several aspects of this progress have been analyzed in some detail by Malerba (1985). First, coinvention has played an important role in many industries. Innovation by sellers and complimentary investments and innovation by buyers in terms of new products and investments in human capital have been very important in many cases. Antonelli (2003) has shown that for the industrial technology sector coinvention involves the technology of the use as well as that of the supplier. User coinvention is especially important in explaining technological change in IT applications, e.g., package software, turnkey solutions, and semi-custom IT solutions. Secondly, the challenge of networks has played a dynamic role in the new innovation technology. The challenge starts from the recognition that innovation and industry evolution are highly affected by the interaction of heterogeneous actors with different knowledge, competences, and specialization. At the general level we know that networks show stability and change over the evolution of an industry. Stable networks are often formed early in the industry life cycle. The empirical evidence shows that major industry-specific events shape the structure of networks. Finally, coevolution plays a most dynamic role in the spread of innovation. In a broad sense coevolutionary processes involve knowledge, technology, actors, demand, and institutions and are often path dependent. Moreover, local learning, interaction among agents, and networks usually generate increasing returns and irreversibilities. Some examples are in order. In pharmaceuticals and biotechnology the interactions between knowledge, technology, and country-specific factors have shaped the evolution of the industry. Another example is telecom equipment and services. The convergence between ICT and broadcasting audiovisual technology and the emergence of the Internet have produced a more fluid market structure, thus expanding the boundaries of the sector by creating new segments and new opportunities. In the software industry the recent spread of networked computing, the development of open system architecture, and the growth of web-based network computing have led to the decline of many large computer producers and also to the emergence of many new specialized software companies. This is a new form of the Schumpeterian process of creative destruction.

Chapter 4
Endogenous Innovation

In Schumpeterian dynamics innovations are endogenous. They are catalytic agents for so spectacular growth and success of the modern capitalist world. They upset the Walrasian general equilibrium theory and its marginalist tradition by introducing several kinds of dynamic shifts over time: new technology, new knowledge, new markets, and new organizational structures. Most of these new dynamic shifts are generated by endogenous forces. Exogeneity is only due to uncertainty whether the outcome is a success or failure in the new venture. The following endogenous aspects are most important:

1. Profit expectations
2. Endogenous growth through innovation
3. Coinvention and coevolution
4. Schumpeterian innovations

We discuss these aspects in some detail. Innovations generally involve new investment in physical or human capital. In Schumpeterian view it includes new forms of organizational improvements, new markets, and coinvention by other firms through technological and knowledge diffusion.

4.1 Profit Expectations

Profit expectations provide the basic motivation for innovation. The Walrasian model of competitive equilibria uses profit maximization as the central objective of firm's input and output strategies. But in equilibrium price equals marginal cost resulting in zero profits, where costs are assumed to include normal profits. Schumpeter considered a dynamic world where this Walrasian equilibrium is altered by the creative accumulation aspect of innovation. Production frontier moves downwards and new profits are generated at higher positive levels. If innovations take the form of knowledge capital or R&D investments, it may lower the unit cost of production through scale economies. Through lowering prices the market may

J. Sengupta, *Theory of Innovation: A New Paradigm of Growth*,
DOI 10.1007/978-3-319-02183-6_4,

expand thus generating still higher profits in a cumulative fashion. If the innovation yields a new product or new process of production e.g., a new medicine or software, it may eventually yield a patent by which quasi-monopoly profits may be earned over 15–20 years. For the period the entry of new firms may be blocked, so that the successful innovator may sustain its strategy for earning super normal profits.

This profit may yield two other effects. One is to induce further investments in new information technology and knowledge. The second is the diffusion effect, which means that other firms may use the information and knowledge available to expand their own investment in related fields. In telecommunications, computer, and pharmaceutical industries this type of diffusion process has promoted a rapid rate of industry growth in recent years. This trend is sure to continue for other industries impacted by the new information technology.

Innovation in the form of R&D investment in knowledge capital has been strongly emphasized in modern theory of endogenous growth. Solow's reliance on neoclassical growth model with perfect competition and exogenous technology cannot fully explain the cross-country variations in per capita income and national growth rates. Romer (1990) and Lucas (1993) have constantly emphasized this point in their endogenous models of growth.

The R&D expenditure process has several distinct features. First of all, there is inherently uncertainty associated with industry-oriented research. Firms investing in R&D process buy themselves a chance at developing a new product or process or software. Newcomers may always enter into similar research activity unless the entry cost is very high. Hence there is always the potential threat of new entry. Firm's expected gains or profits are derived from the probability of success times the market value of the new product or new process. The race for R&D activity between a new comer and the incumbent firm yields an expected monopoly profit for the winner. This has been usually modeled as a Poisson process. For example, if successful R&D innovations arise as a result of a Poisson process with intensity u, the probability that a firm innovates successfully during a period $\mathrm{d}t$ is $u\mathrm{d}t$. If there exist n firms with each firm having an equal chance of winning the R&D race, then the probability of any particular firm becoming a winner in the race is $(u/n)\mathrm{d}t$. The firm's instantaneous expected profit is

$$\pi(n, u) = run^{-1} - c(u, f)$$

where r is monopoly surplus or return and $c(u, f)$ the cost of R&D research with f as fixed cost. Maximization of this expected profit subject to the threat of potential entry provides the incentive for endogenous innovations through R&D investments. By assuming the cost function $c(u, f)$ to be a convex power function in intensity u, Fölster and Trofimov (1997) maximized the profit function with respect to intensity u and this resulted in an optimal profit function to be S-shaped. This type of profit function implies that the positive external impact of R&D innovations sometimes dominates the negative impact from increased competition for the race. They used empirical data of 82 R&D projects for 39 selected Swedish firms during 1988 and

1990 to test the S-shaped profit function. The empirical estimations support the notion of an S-shaped profit curve and they indicate that education, technological competence, and previous sales increase the chance of being the base for new innovating firms.

4.2 Englmann's Model

In contrast with the hypothesis of expected profit maximization, Englmann (1994) considered a disequilibrium model of endogenous innovation and growth. We discuss this model in some detail here. In this endogenous innovation model behavior of agents is assumed to be governed by routines, not by maximization of profits. The entrepreneurs are assumed to invest a fraction of their profits to capital accumulation and another fraction to R&D projects. The latter leads to an increase in labor productivity through an R&D production frontier. Following the Schumpeterian theory of economic development the economic impact of technical change is considered a dynamic disequilibrium phenomenon. Hence steady-state values are not that important. What is important is the time average. We discuss here in some detail Englmann's model of endogenous innovation which includes the diffusion process.

We consider first the R&D process and market entry and then the diffusion process in Englmann's model. A fraction of operating profits allocated to R&D ($s^R P$) is used to employ labor in the research laboratories (L^{R}):

$$L^T = s^R P(1/w) \tag{4.1}$$

where P is profits and w is the wage rate. This labor is assumed to improve labor productivity of the future technology (d_i^R) according to the following logistic learning curve:

$$d_i^R = d_i^R(d_i^{R_{\max}} - d_i^R)\beta_i(L^{\mathrm{R}}/L^{\mathrm{S}}), i = 1, 2, \ldots \tag{4.2}$$

Labor supply (L^{S}) is assumed to grow exogenously at the natural rate n

$$L^{\mathrm{S}} = L^{\mathrm{S}}(0)e^{nt} \tag{4.3}$$

A separate learning curve is assumed for each paradigm i. Here $d_i^{R_{\max}}$ denotes the maximum level of labor productivity achievable within the ith paradigm. Englmann does not consider increasing returns to adoption during the diffusion process and he assumes a deterministic framework. Here the entrepreneurs are assumed to introduce new technology only when the R&D process has led to a higher labor productivity than that of the previous paradigm. This model does not assume the perfect foresight condition. Instead it stipulates a simple decision rule for the market entry of new technology. It will be introduced if

$$d_{i+1}^{R} \geq \epsilon_i d_i, i = 1, 2, \ldots \tag{4.4}$$

Here d_i denotes the actual labor productivity of paradigm i and ϵ_i is a factor such that $(\epsilon_i - 1)$ is the labor productivity differential between the new and previous technology.

For the diffusion process Englmann assumes a Leontief-type production function

$$X_i = \min(d_i L_i, (c_i)^{-1} K_i), i = 1, 2, \ldots \tag{4.5}$$

Here X_i denotes the output produced with the ith technology, L_i the labor input, K_i the capital input, d_i the productivity, and c_i is the capital–output ratio assumed to be identical for different technologies. The prices are assumed to be the same and set equal to one for simplicity. New capital goods can be produced using the same capital goods. With respect to diffusion it is assumed that profits are reinvested in the same technology they were made with. Profits are

$$P_i = \left(1 - \frac{w}{d_i}\right) \frac{K_i}{c_i}, i = 1, 2, \ldots$$

and overall profits are

$$\pi = \sum P_i$$

Two classes of economic agents are assumed, e.g., capitalists and workers. The latter spend what they get with no investment. The capitalists invest part of their profits either in real capital or in R&D. The capitalists' propensity to save s is

$$s = s^K + s^R$$

made up of investment in real capital (s^K) and in R&D (s^R). For simplicity these are assumed to be identical for all technologies. Thus the equations for real capital accumulation may be written as

$$\dot{K}_i = \frac{\mathrm{d}K_i}{\mathrm{d}t} = s^K \left(1 - \frac{w}{d_i}\right) \frac{1}{c_i} K_i, \text{ if } w \leq d_i$$

$$\dot{K}_i \left(1 - \frac{w}{d_i}\right) \frac{1}{c_i} K_i, \text{ if } w > d_i \text{ (scrapping)}$$

The dynamics of R&D employment depending on the wage rate (w) is borrowed from Goodwin's (1967) growth cycle model as

$$\dot{w} = (-m + \ell v)w; m, \ell > 0$$

where v denotes the rate of overall employment and dot denotes a time derivative. Here

$$v = (L^{\mathrm{S}})^{-1}\left(\sum_i L_i + L^{\mathrm{R}}\right)$$

$$L_c = (d_i c_i)^{-1} K_i, i = 1, 2, \ldots$$

Denoting $k_i = K_i/L^{\mathrm{S}}$ as the capital labor ratio the following diffusion equations can be derived as

$$\dot{k}_i = \left[s_i^K\left(1 - \frac{w}{d_i}\right)\frac{1}{c_i} - n\right]k_i \text{ if } w \leq d_i$$

$$\dot{k}_i = \left[\left(1 - \frac{w}{d_i}\right)\frac{1}{c_i} - n\right]k_i \text{ if } w > d_i, i = 1, 2, \ldots$$

These equations are combined with the bargaining equation

$$\dot{w} = \left\{-m + \ell\left[\sum_i \frac{k_c}{d_i c_i} + s^R \sum_{i'}\left(\frac{1}{w} - \frac{1}{d_{i'}}\right)\frac{k_{i'}}{c_{i'}}\right]\right\} w_i$$

where prime denotes the new technology with increased labor productivity.

Three cases of the behavior of this model are distinguished:

Case A: Only one technology is used in production.
Case B: One old and one new (diffusing) technology are used.
Case C: More than two technologies are used.

For case A, only one technology is used in production, the other being developed in the research labs. In this case one can derive the following evolutions of evolution

$$\dot{u}_1 = u_1(-m + \ell v)$$

$$v = \left[1 + (s^R)^{-1}\frac{u_1}{1 - u_1}\right]v^R$$

$$\dot{u}_1 = u_1\left[-m + v^R\ell\left(1 + \frac{u_1}{s^R(1 - u_1)}\right)\right]$$

where $u_1 = w/d_1$. The economically meaningful steady state can be derived as

$$u_1^* = 1 - (s - s^R)^{-1}(nc_1)$$

$$v^{R*} = (m/\ell)\left[\left(1 - \frac{nc_1}{s - s^R}\right)(1 - s^R) + s^R\right]^{-1}(s^R nc_1/s - s^R)$$

This steady state is locally asymptotic stable under normal economic conditions and this state is characterized by the following rate of profit on real capital:

$$r^* = n(s - s^R) \tag{4.6}$$

Thus if the rate of change of labor supply n is equal to zero or negative, the rate of profit vanishes or becomes negative and hence the R&D processes are stopped until a new and more efficient technology is introduced in time.

In case B we have the diffusion of one technology. If the new technology is successful in being more efficient, it crowds out the old technology. The condition for a successful introduction of the new technology implies that the rate of profit earned with the new technology is higher than the old. This leads to an increase in the overall rate of profit which continues during the initial phase of the diffusion process. Then in the final phase of the diffusion process the overall rate of profit tends to fall to its steady-state value r^* of case A, until a new technology is introduced.

For case C we have the diffusion of various technologies. If we define the market shares (x_i) of technology i as

$$x_i = X_i / X_T, X_T = \sum X_i$$

we get for the rate of change of the market share $\hat{x}_i = \dot{x}_i / x_i$.

$$\hat{x}_i = (s - s^R)(r_i - r), i = 1, 2, \ldots \tag{4.7}$$

where r is the overall rate of profit on real capital.

Several comments are in order. First of all, (4.7) implies that not only the most technology can increase its market share but also those whose efficiency is above average. Also the higher the accumulative rate $(s - s^R)$, the higher the change in market share. Secondly, the rate of change of average labor productivity $\hat{d} = \dot{d}/d$, $d = \sum_i d_i x_i$ can be derived as

$$\hat{d} = \frac{(s - s^R)w}{c \sum_i \frac{x_i}{d_i}} \left[\sum_i \frac{x_i}{d_i^2} - \left(\sum_i \frac{x_i}{d_i} \right)^2 \right]$$

where c is the constant capital–output ratio assumed to be identical for all technologies. It is apparent that for a given mean of the labor–output ratio, the rate of change of average labor productivity increases with the variance of the labor output ratio. Hence the more technologies are in use, the higher the rate of technical progress. Finally, several simulations are performed by Englmann for this dynamic model and it is found that a positive rate of skilled labor supply growth is a necessary condition for the economically meaningful steady state or stable limit cycle to exist. This destabilizing tendency can be overcome by sufficiently high R&D and savings rate and a sufficiently high productivity in R&D. This emphasizes the continual need to innovate for sustaining a steady industrial growth driven by the profit motives.

The Schumpeterian model thrives under the competition of new and innovative technologies and the model of hypercompetition emphasized in recent managerial literature provides an operational framework for dynamic analysis.

4.3 Coinvention and Coevolution

Coinvention refers to the interdependence of one invention to another. This interdependence occurs through knowledge spillover and coevolution of modern research especially in telecommunications and software research. Block et al. (2012) have discussed in details the knowledge spillover theory of entrepreneurship and its impact on coinvention and coevolution. Although collaboration in the product market is prohibited by antitrust laws and reputations in most countries, collaboration and cooperation in research are actively encouraged in most countries, and the European Economic Commission has fostered this cooperation actively to encourage intensive research in order to internalize the external economies and spillover effects of modern invention and research development.

We discuss here briefly the knowledge spillover theory of entrepreneurship due to Block et. al. Knowledge spillovers provide an important source of entrepreneurial opportunities called endogenous entrepreneurship. Due to the noncompetitive nature of knowledge as an asset, it may spillover to other firms so that the original producers of knowledge are not able to appropriate the entire value of their knowledge for themselves. These spillovers may then serve as a source of opportunities for other firms and other researchers who may want to start their own enterprises. This frequently happens in pharmaceutical drugs and software products for the telecommunication industries. The knowledge spillover theory of entrepreneurship upholds that the entrepreneurial activity is greater when there is greater investment in knowledge. Spence (1984) has developed models where state subsidies are recommended for promoting research collaboration so that the total research investment for innovation can be substantially expanded.

Coevolution and coinvention enter the very heart of the dynamic analysis of innovation and the evolution of industries and structural change. These coevolutionary processes are sector-specific. In sectors characterized by a system product and heterogeneous demand, coevolution leads to the emergence of a dominant design and industrial concentration. But in sectors characterized by consumers with homogeneous demand, or competing technologies with lock-ins, specialized products and a more fragmented market structure may emerge. In computer industry coevolutionary and coinvention processes firms and their strategies have differed greatly in mainframes, minicomputers, and computer networks. In pharmaceutical and biotechnology the intense interaction between technology knowledge and country-specific factors has shaped the evolution of industry. In software the spread of networked computing, the Internet, embedded software, and the growth of web-based network computing have led to the decline of large computer producers as developers of integrated hardware and the emergence of many specialized software companies.

A very interesting case of coevolution can be found in vertically related industries. Jacobides and Winter (2005) explain the coevolutionary processes between vertically related industries in terms of the coevolution of firm's capabilities and transaction costs and focus their analysis on four factors: knowledge accumulation, capability differences, selection processes, and endogenous transaction costs.

We may now return to the knowledge spillover theory of entrepreneurship developed by Acs et al. (2009). This theory analyzes the effect of entrepreneurship in turning knowledge into different innovation outcomes. Innovation relates to two interrelated processes: the production of knowledge and the exploitation of knowledge. The commercialization of knowledge, especially new knowledge, includes efforts such as financing product development and market research. A high rate of entrepreneurship and exposure to an entrepreneurial climate facilitates the process of turning knowledge into innovative products. Block et al. (2012) used empirical data of 57 firms from 21 European countries from the OECD Economic Outlook data set and the Community Innovative Survey (1996–1998) to run pooled OLS regressions of new-to-the-firm innovation. Their results confirm the proposition that entrepreneurship moderates the relationship between knowledge and the new innovations. These regression results further show that countries with a high rate of entrepreneurship perform better in terms of what is often referred to as 'true' innovation. These findings are consistent with the Schumpeterian view of entrepreneurship and innovation. Schumpeter's theory divided the creative process of economic development into three stages: invention, innovation (commercialization), and imitation. In his view entrepreneurs are innovators who introduce new products, create new production methods, and open new markets. Entrepreneurs are in Schumpeter's view agents who can cope with uncertainty. The entrepreneur's capacity for risk-taking and absorbing uncertainty is more important with *new-to-the-market innovation* than with the new-to-the-firm innovation. The moderating role of entrepreneurship in new-to-the-market innovation is related to the level of product complexity in the manufacturing sector. With increasing complexity of the product and technical production processes, imitation is hardly a feasible way for new firms to enter the modern manufacturing sector. A firm's chances of survival increase when it has 'true' innovations, i.e., innovations that generate new products, new production methods, or new markets.

4.4 Schumpeterian Perspectives

The relationship between innovation and industrial change has always been central in Schumpeter's work in various ways. He was very much interested in innovation either as a process of creative destruction or as a process of creative accumulation. Both processes are endogenous in the sense that the profit motive is central to the entrepreneurial strategies here. At the modeling level both processes use models of industry dynamics with rational actors and technological learning by incumbents or entrants or both, and the competitive process weds out the heterogeneity in innovation, and industry evolution is linked to structural change and the changing sectoral composition of the economy as a whole.

Several formulations of Schumpeterian dynamics are available in the literature. We discuss two here. One is by Aghion and Howitt (1998) and the other by Palokangas (2007). The first formulation abstracts from capital accumulation completely. Households with N individuals maximize the utility function $u(y) = \int_0^\infty \exp(-rt)y(t)dt$. The output y of the consumption good depends on the intermediate good x according to the log linear production function $y = Ax^\alpha$, $0 < \alpha < 1$. Innovations are assumed to consist of the invention of a new variety of intermediate good that replaces the old one. This replacement process improves the technology parameter A by a constant positive factor g. Society's fixed stock of labor $L = x + n$, where x is the amount used in manufacturing and n the amount used in research. Innovations come in a sequence and the mth innovator maximizes the profit flow π_m by choosing x_m:

$$\pi_m = \max_x [p_m(x)x - w_m x]$$

where w_m and $p_m(x)$ are wage and price. Assuming a competitive market for the final good with an inverse demand function $p_m = A_m \alpha x^{\alpha-1}$ for the mth innovator, the first order conditions yield the optimal values x_m^* and π_m^* as

$$x_m^* = \left(\frac{\alpha^2}{w_m/A_m}\right)^{1/(1-\alpha)}$$

$$\pi_m^* = \left(\frac{1}{\alpha} - 1\right) w_m x_m^*$$

Note that both x_m^* and π_m^* are decreasing functions of the productivity adjusted wage rate (w_m/A_m). Thus π_n^* decreases with respect to (w_m/A_m) due first to the creative destruction and then for the negative dependence of current research on the amount of expected future research. Specifically, a higher demand for future research labor will push wages w_{m+1} up, thereby decreasing the flow of profits π_{m+1}^*. This in turn will tend to discourage current research, i.e., to drive n_m down. Also, one can derive from the steady-state equilibrium two important conclusions. One is that the steady-state level of research $\bar{n}$ is a decreasing function of α. Second, the more the intensity of competition, the lower the monopoly rents which will be appropriated by the successful innovators. This process will yield smaller incentives to innovate. Several comments are in order for this model. First, the cost-reducing aspect of innovation is not analyzed in this framework. Yet the success of a new innovator depends very critically on this aspect of efficiency. Caballero and Jaffe (1993) developed a new methodology for measuring research productivity through innovation. They assessed the extent of Schumpeterian creative destruction, knowledge obsolescence, and knowledge spillovers in the endogenous growth process using US data on patents and patent citations. They find an average annual rate of creative destruction of between 2 and 7 percent during the 1970s with rates up to 25 percent in the pharmaceutical sector. They used the following equation for describing the growth of general knowledge:

$$\dot{A} = \frac{\mathrm{d}A}{\mathrm{d}t} = \Gamma \lambda n_t$$

where Γ denotes the current state of knowledge, n_t is the current flow of research labor, and λ is the parameter indicating the productivity of research technology. The knowledge variable Γ is taken to be equal to the integral

$$\int_{-\infty}^{t} a(t,s)\lambda n_s \mathrm{d}s$$

where

$$a(t,s) = \delta e^{-\beta(A_t - A_s)} \left(1 - e^{-\gamma(t-s)}\right)$$

is the marginal contribution of a vintage s sector to current innovations, β is the rate at which an old idea becomes obsolete, and γ is the rate of diffusion of older ideas into current general knowledge. This marginal contribution in turn is taken to be proportional to the patent citation rate

$$a^*(t,s) = C_{t,s}/(S_t P_s)$$

where $C_{t,s}$ is the number of observed citations by current patents of year t of older patents by year s, P_s is the total number of patents of year s, $s < t$, and S_t is the number of sample patents in current year t. On the assumption that a^* is proportional to a Caballero and Jaffe estimate parameters that allow them to characterize the process of creative destruction. Some of the important findings are as follows:

1. Ideas diffuse rapidly within one or two years.
2. The rate of ideas' obsolescence measured by the parameter β above has increased from 3 % at the beginning of the century up to 12 % in 1990, and it has averaged around 7 % between 1975 and 1992.
3. The ratio $(\dot{A}/n)$ where $\dot{A}$ is measured by the number of new innovations has decreased substantially between 1960 and 1990 by approximately 30 %.

Aghion and Howitt (1998) have analyzed in some detail the interrelation between the long-run national economic growth and several endogenous growth models and found that the long-run rate of growth is positively correlated with the flow of patents, the flow of entry of new firms, and the flow of new product innovations.

In their view the central role of creative destruction in Schumpeterian growth theory can be empirically tested by analyzing the correlation between growth and two other variables: the flow of exit of firms and the rate of obsolescence of capital. The former is identical to the flow of entry in a steady-state growth equilibrium and the latter is equal to the rate of arrival of new innovations.

Palokangas (2007) emphasized a different approach and developed a stochastic model of creative destruction. In this formulation the new innovations replace the old and the innovative firms generate a stream of productivity improvements for the industry. Varieties of innovations occur through multiple new products and services and also multiple intermediate goods. We discuss in brief this stochastic model. Here each innovative firm produces output Y_j by using labor L_j and capital K_j. The productivity of labor is assumed to be unity in R&D and α in other production activity. R&D investment I_j uses labor z_j and it is assumed that there is no depreciation in R&D investment. The production function assumes constant returns to scale as follows:

$$Y_j = F(aL_j, K_j) = f(\ell_j)K_j$$
$$\ell_j = aL_j/K_j,\ f' > 0,\ f'' < 0$$

where prime denotes partial derivative and the firm's budget constraint is

$$C_j + w_j L_j + v_j Z_j = A^{g_j}(Y_j - I_j)$$

Each firm produces the same consumption good C_j and a firm-specific capital good K_j. Consumption good C_j is produced by converting the residual output $(Y_j - I_j)$ in proportion A^{g_j}, where $A > 1$ is a constant and g_j is technology. The wage costs are $w_j L_j$ and R&D costs are $v_j Z_j$, which are assumed exogenous. The R&D innovation is stochastic and it is assumed to follow a Poisson process with the probability of success $(\lambda \log z_j)\mathrm{d}t$, z_j begin equal to Z_j/K_j. A success leads to a new technology, new product, or a new market. By combining this stochastic process with the production function he derives a stochastic differential equation for the capital accumulation process as follows:

$$\mathrm{d}K_j = I_j \mathrm{d}t = \left\{f(\ell_j) - w_j L_j\right\} K_j - A^{-g_j}(c_j + v_j z_j)$$
$$\text{with } c_j = C_j/K_j$$

On maximizing the expected value of a utility function $U_j = (1-\sigma)^{-1}\,[c(t)^{1-\sigma} - 1]$ as

$$\max J = E\int_0^\infty e^{-\rho(\tau - t)} U_j \mathrm{d}\tau$$

One may derive the optimal paths of c_j, ℓ_j, and z_j. Several implications of the model are important for analysis. First, the model assumes innovation shock and technology as endogenous as they depend on the innovative firm's optimal decision-making process. Second, the model assumes that a successful development of new technology generates an upgrading of the previous vintage capital which yields higher industry productivity and lower unit costs and prices. Several comments may be made about this model. One is that the model assumes a competitive framework of the market. But innovations frequently involve rivalrous competition and the dominance of the successful innovators. Secondly, the spillover effects and externalities are not at all considered here. Even patent citations provide some noisy signals of the presence of spillovers and the aggregate citation flows and networks provide a useful proxy for knowledge spillover intensity. Some authors like Spence (1984) developed models where spillover effects act as deterrents to new R&D investments by innovating firms. Also the market structure assumed by Spence is the Cournot–Nash framework instead of a procompetitive form. Finally, complementarities flowing from innovations are often a major source of increasing returns. This aspect needs more emphasis in this model.

Schumpeter's model of creative destruction may be viewed in two ways. One is the creative aspect and the other the diffusion aspect. Creative accumulation of knowledge capital and creative ventures involve substantial R&D investment. This involves significant risks, but the profit expectations are also very high. High profit incentives keep firms searching for ways to explore opportunities through new innovative activities.

4.5 Public Research Expenditure

Public expenditure on R&D occurs through government, universities, and nonprofit research organizations. These expenditures help industry growth in many ways. Though profit incentives are not behind these expenditures, such R&D spendings are market oriented in many ways. For many countries such as China, Taiwan, and South Korea, state-sponsored research resulting in new products or new processes is eventually privatized, so that the private industries can develop them further. Secondly, state subsidies to nonprofit private research organizations are in many ways directed to those industries, where private entrepreneurs are less inclined to invest. These are typically pharmaceutical industries and infrastructure-related industries. The spillover effects from these industries to the private sector are very large. Finally, the creation of knowledge and its diffusion through university research help the growth of learning-by-doing effect and the professors and research personnel involved in state-sponsored basic research frequently migrate to the private sector directly or indirectly. In many European countries private industries form research collaboration to internalize the spillover effects. This is directly allowed by the law, and the state frequently provides subsidy to this collaborative ventures. The newly industrialized countries (NICs) of Southeast Asia which have maintained high growth rates over the last three decades have consistently adopted this policy for lowering the unit production costs and improving their dynamic efficiency.

We discuss now a few theoretical models relating public research expenditures to overall economic growth with industry growth as follows:

A. State subsidy model due to Spence
B. Public goods model due to Barro
C. Learning by doing and secondary innovations
D. Model of technical infrastructure expenditure

A. The Spence Model

R&D investments through innovation are intended to reduce costs and frequently these investments take the form of developing new products that reduce price and hence expand demand and new markets. Two sorts of problems arise in this framework in industry performance. Cost-reducing expenditures are largely fixed costs, sometimes with negligible marginal costs. Profitability or revenues are the goal of R&D investments, but revenues understate the social benefits both in the aggregate and at the margin. Hence in optimality of outcomes results here. Secondly, the market structure in this environment is most likely to be concentrated and imperfectly competitive. This may result in a decline of allocative efficiency. Added to this framework is the appropriability problem, which is sometimes referred to as externality problems. If the R&D for the single firm is not appropriable, the initial incentives to do the R&D are reduced. Restoring appropriability is sometimes regarded as a second best solution to the incentives problem because it creates monopoly power. Thus there appears to be a trade-off between incentives for innovative R&D on the one hand and the economic efficiency with which the industry achieves its level of cost reduction.

We follow Spence in the model formulation where $c_i(t)$ is unit costs for firm $i = 1, 2, \ldots, n$. Unit costs depend on the accumulated knowledge in the industry obtained by firm i:

$$c_i(t) = F(z_i(t))$$

where $F(z_i)$ is a decreasing function of z_i. Let $m_i(t)$ be the current R&D expenditure by firm i, where

$$\dot{z}_i(t) = \mathrm{d}z_i(t)/\mathrm{d}t = m_i(t) + \theta \sum_{j \neq i} m_j(t)$$

Here the parameter θ captures spillovers. Thus if $\theta 0$, there are no spillovers or externalities. If $\theta = 1$, the benefits of each firm's R&D are shared completely. For the normal case $0 < \theta < 1$, which suggests that spillovers are imperfect. The benefits from the sale of x units of the good are $B(x)$, with the inverse demand function denoted by $B'(x)$. The profits of firm i are

$$E^i = x_i B'(x) - c_i x_i, \; x = \sum x_i$$

Spence assumes that there is a Nash equilibrium at each point of time in the market that depends on the costs $c = (F(z_1), \ldots, F(z_n))$ or on $z = (z_1, \ldots, z_n)$. Here we can think of $E^i = E^i(z)$ as the present value of firm i's earnings gross of R&D investment. Assume there is a subsidy s for R&D so that the firm's net R&D costs are $(1 - s)$ dollars. Let $M_i = \int_0^\infty m(\tau)\mathrm{d}\tau$ be the accumulated investment in R&D where

$$z_i = M_i + \theta \sum_{j \neq i} M_j$$

The firm's net present value of its earnings can be written as

$$V^i = E^i(z) - (1 - s)M_i$$

Following the Cournot–Nash framework the firm i takes the M_j of its rivals as given and maximizes V^i with respect to M_i by setting the equation

$$E_i^i + \theta \sum_{j \neq i} E_j^i = 1 - s$$

where E_j^i denotes the derivative of E^i with respect to z_j. The solution to these n equations yields the market equilibrium. In the symmetric case the level of z_j will be the same for all firms denoted by z. The market result in z in equilibrium is the maximum of

$$R(z) - (1 - s)z$$

where R can be viewed in the symmetric case $M_i = M$ as

$$R'(v) = 1 - s$$

Thus the market acts as if it were maximizing $[R(v) - (1 - s)v]$ with respect to v where

$$v = [1 + \theta(n - 1)]M$$

The function $R = R(z, n, \theta)$ captures the market incentives with respect to R&D investment. On assuming $E_j^i < 0$, $i \neq j$ one can see that $R_\theta = \partial R/\partial\theta < 0$ and $R_{z\theta} = \frac{\partial}{\partial\theta}\left(\frac{\partial R}{\partial z}\right) < 0$. Thus an increase in spillovers reduces the incentives for R&D and the cost reduction and will reduce the amount of cost reduction in market equilibrium. Furthermore, R&D costs at the industry level can be written as

$$\text{R\&D} = zn/[1 + \theta(n - 1)]$$

This shows that for $n > 1$, the R&D costs of the achieved amount of cost reduction declines as θ increases with $\theta > 0$; the unit costs have an upper limit of $1/\theta$ as n increases. Thus while spillovers reduce the incentives for cost reduction, they also reduce the costs at the industry level of achieving a given level of cost reduction. The incentives can be restored through state subsidies. Thus it follows that spillovers can improve overall performance of the market with the incentives appropriately restored by state subsidies.

Two important comments have been made by Spence. One is that the provision of state subsidies to restore market incentives has the added benefit of lowering entry barriers, increasing competition and improving allocative efficiency. Secondly, the R&D output has the character of a public good. The incentives are weak for individuals to supply it. But we do not generally approach the solution by compelling the beneficiaries to pay for it where possible because that leads to underconsumption and suboptimal use. It is preferable to supply the public good publicly or subsidize the private supplier.

B. Public Goods Model of Barro

This model assumes that government expenditure (G) is a public good. It is nonrival and nonexcludable. Hence each firm makes use of all of G and one firm's use does not diminish the quantity available to others. Let $\hat{G} = aG$ be the public expenditure on research, e.g., the NSF, NIH, and other research fundings of universities and nonprofit research agencies which help the growth of knowledge. This knowledge creation and diffusion help stimulate the endogenous growth of the whole economy. We follow Barro (1990) in assuming a Cobb–Douglas production function for each firm i as

$$Y_i = AL_i^{1-\alpha}K_i^{\alpha}(\hat{G}/a)^{1-\alpha}, 0 < \alpha < 1$$

Here each firm exhibits constant returns to scale for private inputs L_i and K_i. For fixed G the economy faces diminishing returns to aggregate capital $K = \sum K_i$. If however G rises with K, then the production function specifies constant returns in K_I and G for fixed L_i. For this reason the economy is capable of endogenous growth, since diminishing returns are avoided. Also G_i are complementary with the private inputs in the sense that an increase in $\hat{G}_i$ raised the marginal production of L_i and K_i.

Denoting τ as the tax rate for $G = \tau Y$, the firm's after-tax profit is

$$L_i[(1-\tau)Ak_j^{\alpha}G^{1-\alpha} - w - (r+\delta)k_i]$$

where $k_i = K_i/L_i$, w is the wage rate, and $(r + \delta)$ is the rental rate. Profit maximization and zero profit condition now imply that the wage rate equals the after-tax marginal product of labor and the rental rate equals the after-tax marginal product of capital. If we set $k_i = k$, then the rental price of capital may be written as

$$r + \delta = (1-\tau)(\partial Y_i/\partial K_i) = \alpha A^{1/\alpha}(\tau L)^{(1-\alpha)/\alpha}(1-\tau)$$

If L and τ are constant, then the after-tax marginal product and hence the rate of return r is invariant with k and increasing with L. Thus an increase in scale represented by L raises the after-tax marginal product of capital and expands the social marginal product in a parallel way. A higher L leads to higher values of the decentralized growth rate.

Two comments are in order. First, the complementarity of public research represented by $\hat{G}_i$ generates increasing returns through spillover effects and this has been an important factor in the rapid growth of NICs emphasized by Romer, Lucas, and others. Secondly, the knowledge created by government in sponsoring research is a serious candidate for a significant public good, since it generates the process of learning by doing which acts as a catalysis for endogenous industry growth.

C. Learning by Doing and Secondary Innovations

Innovations are diverse in form and this diversity cannot be captured by an aggregate model of endogenous growth for the whole economy. Aghion and Howitt (1998) have discussed in detail this aspect of innovation diversity. We follow Chap. 6 of his book: *Endogenous Growth Theory* for a review of their analysis below.

Any new product, new technique of production, or new market is created not by one innovation but by a whole sequence of innovations. Some innovations are more fundamental than others in the sense that they open up new windows of opportunity for future development. Information technology in the form of iPhones, iPads, and other communication devices belongs to this category. Other innovations are secondary. Fundamental and secondary innovations in the form of research knowledge are complementary and nonrival activities. Aghion and Howitt have noted three important characteristics of fundamental research and related innovations. First, this knowledge can be generated by learning by doing as well as by research. Secondly, the level of research and the rate of overall economic growth can be increased if production works become more adaptable and flexible in the sense that they can easily switch from producing old products to producing new ones. Flexible manufacturing processes thus occupy a central role in stimulating new industry growth. This result supports Lucas's (1993) claim that the key to success of the NICs in Asia in rapid economic growth for the last three decades is their ability to move skilled workers quickly between sectors. Finally, any parameter change that raises the productivity of the innovation process will shift resources out of learning by doing into productive research. This is because research capitalizes more of the benefits of increased growth.

We consider briefly the model of endogenous innovation due to Aghion and Howitt. The model assumes H skilled workers, where each can choose whether to engage in research or production. There is a single final good which can be used only as a consumption good and a continuum of intermediate goods which constitute the only inputs into the final good. The production matrix has the following characteristics:

1. Final output is produced with a continuum of intermediate goods of different vintages.
2. If H^r denotes workers using general knowledge to do research, then the flow of new products is $\lambda^r H^r$ where λ^r is each researcher's Poisson arrival rate of *fundamental innovations*.

3. Intermediate good of more recent vintages are potentially better because they embody a higher level of general knowledge. The aggregate final output at date t is

$$Y_t = \int_{-\infty}^{t} \lambda^r H^r A_\tau Z_{t-\tau}(x_t - \tau)^\alpha \mathrm{d}\tau, 0 < \alpha < 1$$
$$= \int_{-\infty}^{t} Y_{t,\tau}\mathrm{d}\tau, z_a = \text{quality of the good}$$

where $Y_{t,\tau}$ is the production function.

4. Quality improvements come at a rate equal to the flow of *secondary innovations* across the whole economy and is denoted by LBD. Learning by doing (LBD) occurs in each firm at the rate of $\lambda^d(x_a)^{1-v}$, where λ^d is the productivity of learning by doing and $0 < v < 1$. Thus production workers produce two joint products: final output and secondary innovations. The firm is assumed to appropriate the output but not the innovations. Hence

$$Z_0 = 0 \text{ and } \frac{\mathrm{d}Z_a}{\mathrm{d}a} = \text{LBD} = \int_0^\infty \lambda^r H^r \lambda^d (x_s)^{1-v}\mathrm{d}s, a > 0$$

5. Finally new general knowledge is created by research and learning by doing throughout the whole economy, using the existing stock of knowledge.

The model assumes that of growth of general knowledge ($\dot{A}/A$) is a function of the current flow of innovations of both types: fundamental and secondary and also the accumulated stock of previous knowledge, i.e.,

$$\dot{A}_t/A_t = G(\lambda^r H^r, \text{LBD})$$

which satisfies:

(a) $G = 0$ when $H^r = 0$ and when LBD $= 0$.
(b) G is strictly increasing and concave in both arguments.

It can be shown that in a steady state the economy's growth rate will equal its growth rate of general knowledge. On combining the equations above we can arrive at the final growth equation

$$g = G\left(\lambda^r H^r, \frac{(\lambda^r)^v \lambda^d}{\sigma^v(1-v)}(H^r)^v(H - H^r)^{1-v}\right)$$

when σ is the Poisson rate of upgrading of a researcher. Thus growth can be expressed as a concave function of the level of research H^r. Clearly the growth rate achieves a maximum value g^* by concavity and till g^* is reached, further increases in research will raise the growth rate, but beyond that point it will reduce the growth rate. This has important implications for public education policy, i.e., an economy that found itself beyond the point of maximum growth g^* would be better off channeling its public support to primary, secondary, and vocational education rather than to higher education and higher research.

To conclude this section, we note that fundamental research opens up new windows of opportunity or new product lines, learning by doing fills up those windows and/or brings incremental improvements to those product lines. Fundamental and secondary innovations are thus complementary.

Finally we note the positive effect of the upgrading rate σ on research and growth. Because the growth curve is upward sloping at the steady-state point this upgrading will raise the growth rate. As long as the growth of fundamental knowledge does not depend too much on the rate of learning, the growth curve will not shift enough to reverse this positive effect. This has been called the *"Lucas Effect"* where adaptability increases growth.

D. Technical Infrastructure Model

In this model we assume that all public expenditure has positive effects on the technical infrastructure of the economy. The infrastructure includes knowledge structure of the industry along with the physical infrastructure in the form of transportation and institutional framework. Alfred Marshall characterized the technical and knowledge infrastructure as conjecture benefits, which reduce the unit costs of the firms. His concept of the particular expenses curve for a given industry specified the heterogeneity of the distribution of firms in an industry and the impact of fundamental innovation is much greater the more heterogeneous is the firm distribution, since the processes of entry and exit may act more quickly.

We consider a simple model of productive government expenditure providing the knowledge capital and enhancing it through information diffusion. We assume the production function for the economy as

$$Y = \alpha\left(\frac{H}{K}\right) K, \alpha' > 0, \alpha'' < 0$$

where prime denotes partial derivative with respect to the argument. Here H denoted government expenditure on technical infrastructure of the economy which raises the marginal product of capital K with diminishing returns as in the neoclassical framework. Abstracting from government consumption, the first best central planning problem is to maximize

$$\int_0^{\infty} (\gamma)^{-1} C^{\gamma} e^{-\beta t} \mathrm{d}t$$

subject to

$$\dot{K}_t = \alpha\left(\frac{H(t)}{K(t)}\right) K(t) - C(t) - H(t)$$

Here C is total consumption expenditure and we assume that the government ties its expenditure to the capital stock

$$h = H/K, 0 > h > \alpha$$

both at the actual level of government expenditure growth rate with the economy. We distinguish two cases. In one h is set arbitrarily and in the other case h is set optimally along with C and K. In the first case the following equilibrium paths emerge when the planner makes the optimal decision:

$$C/K = (1-\gamma)^{-1}[\beta - \gamma(\alpha(h) - h)]$$
$$\dot{K}/K = (1-\gamma)^{-1}[\alpha(h) - h - \beta] = \phi \text{ say.}$$

If h is set optimally along with C and K, we have to impose the additional optimality condition

$$\alpha'(h) = 1$$

This implies that the optimal ratio h^* of infrastructure to capital is attained where the marginal benefits to productivity just match the unit resource costs of additional infrastructure spendings by government. It is easy to see that

$$\partial\phi/\partial h = (1-\gamma)(\alpha'(h) - 1)$$

which implies that the equilibrium growth rate is maximized at the optimal infrastructure–capital ratio h^*.

Two comments are in order. First, if we assume increasing returns in the production function, our framework is altered and we have to introduce endogenous growth as we have done in earlier sections. Secondly, technical infra may also be viewed in terms of the quality of human capital as Lucas and others have done.

4.6 Endogenous Industry Growth

Innovation is central to modern industry growth and industry growth generates endogenous growth for the whole economy. Innovations have diversity. Heterogeneity of firms in an industry yields this diversity and the diffusion process and externalities spread this diversity further. Innovations have several dynamic characteristics. First, they have direct impact on production costs and efficiency. This occurs through upward shifts in the firm's production frontier. Frequently this involves a race in R&D investments among competing firms. It also leads to quality improvement in existing goods and services. For example, a pharmaceutical firm develops an improved drug through R&D investment over several years. It then becomes the new leader, the winner of the R&D race. It raises price exactly to the extent of quality improvement. At this price the leading firm becomes a monopoly producer with quasi-monopoly profits. Sometimes the patent for the new product may protect the monopoly market for 20 to 25 years. A second dynamic aspect of innovation is the process of routinization of innovations in oligopolistic or rivalrous competition and the spread of incremental innovations. The latter involves industry-wide transmission of new technology and its cumulative multiplier effects. Baumol (2002) has considered this process as the dynamic engine of unprecedented growth of the capitalist economy in modern times. Finally, increasing productivity through innovations raises competitiveness in industry performance and Porter (2012) has argued that the USA has amazing core strengths whereby the workers can improve productivity and reduce unit costs. He found out in his survey report that about 71% of Harvard Business School alumni think that US competition is declining and this poses some risks to the continuing prosperity. More innovations and R&D investments and improvements in education are called for.

Theoretically innovation has its greatest impact on the information content of production, distribution, and the markets. The competitive paradigm fails to capture this information explosion, which has alerted us that equilibrium may not matter due to market failure and scale economies. History matters and the forward-looking view are more important for growth. Equilibrium processes are less suitable for analyzing the growth dynamics than the evolutionary processes.

Sengupta (2012) has recently discussed the various endogenous aspects of industry growth which have long-run impacts on the overall economic growth. Modern economies today have undergone a dramatic change, thanks to the advent of the personal computer and communication technology. There has been a dramatic shift from material manufacturing to new innovation technology with R&D and human capital. We have entered a new information age, where efficient channels of information usage in all modern industries have achieved substantial productivity gains through increasing returns processes, learning by doing, and incremental innovations. These have generated catalytic forces for endogenous growth and its spillover across the world.

Chapter 5
Innovation Diversity: Industry Applications

Innovation is a broader concept than technology. It includes technology, knowledge capital, and organizational competence. In modern industries the information content of innovations has played a critical role in industry growth. Innovation comes in varieties. It may take several forms. Some of the most important forms are as follows:

1. Fundamental vs. secondary innovations
2. Specific vs. general purpose technology
3. Product vs. process innovations
4. Endogenous innovations
5. Knowledge-based innovations

As for specific industries the following comprises some of the most important areas of application:

A. Information and communication technologies (ICT)
B. Pharmaceutical industry
C. Innovation in services
D. Telecom industry: structure and growth

5.1 Innovation Types

Fundamental innovations are nonroutinized. They can be viewed as dynamic shocks to the static Walrasian equilibria. Secondary innovations oftentimes introduce routinized forms, which help reduce uncertainty. Baumol (2002) has discussed three growth-creating properties of fundamental innovations as follows:

1. The cumulative character of many independent innovations creates new technical knowledge. This enhances the spillover effect and the economy's store of knowledge capital.

J. Sengupta, *Theory of Innovation: A New Paradigm of Growth*,
DOI 10.1007/978-3-319-02183-6_5,

2. The public good property of such innovations increases the economies of scope in the generation of technological improvements for a multiplicity of firms. However, this property also has the adverse effect that it leads to nonoptimal levels of innovation investment. Appropriate stat policy is therefore needed here for correcting the inoptimality.
3. These innovations also generate accelerator effects on other innovations through backward and forward linkages.

Specific purpose technologies are processes of incremental innovations, which are different from general purpose technologies (GPT). GPT are more flexible and they have significant scale effects. Very often GPT is driven by product quality. A large economy produces more varieties of products; hence this requires spreading the quality improving R&D expenditure over a wider range of products. Growth in software technology has made it much easier for GPT to be applied on a wider scale. SPT is very frequent in the pharmaceutical industries, where specific drugs are developed for specific diseases, where GPT is more prominent in manufacturing, and where flexible manufacturing has recently expanded the range of innovations on a wide scale. Over the last two decades shorter product life cycle and hypercompetition have forced many firms to adopt a higher degree of automated production and distribution processes. Their objectives are to improve both flexibility and productivity. A study of the US Dept. of Commerce estimates that over the 14-year period 1981–1995 the number of CAD (computer-aided design) installations increased 40-fold in the USA from the level of 4,500 in 1981. According to recent estimates of the European Commission of Europe the world investment market for flexible industrial automation has exceeded 2.5 trillion dollars in 2005. All these massive investments are in one form or other investments in manufacturing flexibility.

One may cite two major reasons why flexibility of a GPT plays a more important role in industry growth. The first reason is the productivity effect due to its capacity to adapt to changing environments in processes and markets. The total system flexibility of an FMS can be viewed as a weighted combination of two factors Q and E, where Q is the speed of response to a shock and E is the economic response in terms of costs to a change factor. Q can be measured for example in terms of lead time between customer's demand and completion of the order. The factor E is a measure of the economic factors such as inventory and current rates of machine utilization. The second reason for flexibility is the *economies of scope*, brought about by fundamental innovations. Whereas the economies of scale reduce unit costs due to size, the economies of scope yield cost savings due to increasing complexity of variety of products. Here innovation as technological progress shows up as an expansion of the number of producer and consumer products.

5.2 Expansion in Variety of Products

The process innovation differs from the product innovation in that it has a long-run cumulative effect. New products have a life cycle of 6 to 7 years, but new processes allow the adaptation and improvement by successive incremental innovations. These successive innovations frequently produce N varieties of intermediate goods. An expansion in the number N requires technological innovation and intensive R&D investments. For endogenous process innovation generating a variety of intermediate goods we assume that the inventor of good j retains a perpetual monopoly right over the production and sale of the good X_j that uses his design or the process. The flow of monopoly rentals will then provide the incentive for innovation. For fixed N we follow Romer (1990) to write the production function for firm i as

$$Y_i = AL_i^{1-\alpha} \sum_{j=1}^{N} (X_{ij})^{\alpha} = AL_i^{1-\alpha} (NX_{ij})^{\alpha} N^{1-\alpha}$$

where we assume $X_{ij} = X_i$ in equilibrium. Suppose that once invented an intermediate good of type j costs one unit of Y to produce. The present value $V(t)$ of returns from discovering the j-th intermediate good is then

$$V(t) = \int_t^{\infty} (P_j - 1) x_j e^{r(v,t)} \mathrm{d}v$$

where X_j is the total quantity of intermediate goods produced at t and $r(v, t)$ is the average interest rate between times t and v. The monopolist sets the price P_j to maximize profits $(P_j - 1)X_j$. This yields the optimal monopoly price as

$$P_j = p = 1/\alpha > 1, \text{given } \alpha < 1$$

Following Romer (1990) we assume there is free entry into the business of being an inventor so that anyone can pay the R&D cost to secure the net present value $V(t)$. If $V(t) > \eta$, then an infinite number of resources would be channeled into $R(t)$ at time t. If $V(t) < \eta$, then no resources would be allocated to R&D. Hence in equilibrium we would want $V(t) = \eta$ for all t. In equilibrium the interest rate equals a constant r and we obtain

$$r = (L/\eta) A^{1/(1-\alpha)} \left(\frac{1-\alpha}{\alpha} \right) \alpha^{2/(1-\alpha)}$$

The underlying technology and market structure peg the rate of return at the value r. The marginal $(N + 1)$th intermediate good generates a present value of monopoly profits that just covers the R&D cost η. Following the endogenous growth theory we assume that households maximize the utility function U over an infinite horizon

$$U = \int_0^{\infty} e^{-\rho t} \left(\frac{c^{1-\theta} - 1}{\theta} \right) \mathrm{d}t$$

Households earn the rate of return n on assets and receive the wage rate w on total labor L. Households optimization leads to the growth rate of consumption γ_c as

$$\gamma_c = (1/\theta)(r - \rho) = \gamma$$

where the same growth rate γ applies to the number of designs N and to output Y and consumption C. For reasonable economic conditions, the above equation is valid if the parameters lead to $\gamma \geq 0$. Hence we assume this nonnegative condition. Several implications now follow. First of all, a greater willingness to save, i.e., lower ρ and θ, and a better technology in the form of higher A raise the growth rate γ. Secondly, a decrease in η raises the rate of return r and therefore raises the growth rate in the equation above. The model also contains a scale effect: a larger labor endowment L raises γ in the equation

$$\gamma = (1/\theta) \left[(L/\eta) A^{1/(1-\alpha)} \left(\frac{1-\alpha}{\alpha} \right) \alpha^{2/(1-\alpha)} - \rho \right]$$

This effect is similar to the model of learning by doing with spillovers. Finally, the expansion in the quantity of intermediate goods x provides a static and dynamic gain in efficiency. In a static framework with fixed N, the monopoly pricing implies that the marginal product of x exceeds its cost of production and therefore the economy fails to maximize the goods available for consumption. If however more outputs are allocated to X, then the consumption could rise. The government subsidy to purchase of x thus allows the economy to secure this static gain.

The higher the level of X also has a dynamic effect that involves the incentive to expand N over time. The increase in the quantity of intermediate goods X raises the flow of monopoly profit by a factor $(1/\alpha)^{1/(1-\alpha)}$. This increase in profit raises the rate of return r by the same factor; hence the private rate of return coincides with the social rate. Note however that the result that both static and dynamic inefficiencies from monopoly can be eliminated by a single policy instrument as a government subsidy depends on the underlying form of the production function. Romer's model of technological change generates endogenous growth because he assumes that the cost of inventing a new product declines as society accumulates more ideas and knowledge proxied by N. Although the growth rate is constant in equilibrium, the determination of this growth rate in a decentralized economy involves a new type of externality: an individual firm's decision to invest in R&D and therefore to expand N reduces the required amount of labor needed for subsequent innovations. Current research therefore has a positive spillover on the productivity of future research.

5.3 Optimal Growth with Endogenous Innovation

Endogenous growth theory implies that with enough innovations of all types and the right direction of innovations sustainable development is possible. We consider here two growth models which embody this theory of sustainable development. One is the AK model and the other the Schumpeterian approach to endogenous industry growth through innovation. In the first case the optimal growth is obtained by maximizing the lifetime utility function

$$U = \int_0^\infty e^{-\rho t} u(c)\mathrm{d}t$$

subject to the constraint

$$\dot{K} = Y - C$$

where $u(c) = (1-a)^{-1}(c^{1-a}-1)$ is the isoelastic utility function with c as per capita consumption. Following the AK model we assume the linear production function

$$Y = AK$$

where K is an aggregate of different sorts of capital goods and also the current state of technological knowledge in the economy and A is a positive constant. It can be shown that the case $A < \rho$ is degenerate since it will then be never optimal to accumulate more K as the marginal product falls short of the rate of time preference. Hence we assume that $A > \rho$. Then it would be optimal to accumulate more capital indefinitely. In this case there exists a steady-state optimal path in which consumption, capital, and output all grow at the same rate g^* where

$$g^* = (1/a)(A - \rho)$$

Since c can be written as $c = (1 - s)y$, where s is the savings rate and y per capita income, the growth rate g^* can be also written as

$$g^* = sA - n$$

where n is the rate of growth of labor. Several comments are in order. First of all, this AK technology can display positive long-run per capita growth without any technological progress. Moreover the per capita growth rate depends on the behavioral parameters of the model such as the saving rate (s) and the rate of population growth (n). Unlike the neoclassical model of Solow, a higher saving rate s leads to a higher rate of long-run per capita growth g^*. Similarly if the level of technology A improves once and for all, then the long-run growth rate is higher. Second, the knowledge base of human capital is included here in K; it may help eliminate the tendency for diminishing returns. Here the learning of technical knowledge by one producer may raise the productivity of others through a process of spillovers of knowledge from one producer to another. Hence a larger industry-wide or economy-wide capital stock improves the level of technology for each producer. Consequently diminishing returns to capital may not apply. Finally, the empirical evidence suggests that while the AK model may not provide a good description of the historical growth process in most growing countries, it cannot be rejected for the economy of China. The econometric tests by Sengupta (2011) show that sustained physical investment activity was indeed associated with China's long-run growth, where growth may have been predominantly driven by investment in heavy industry and services.

We now consider the Schumpeterian approach as discussed by Aghion and Howitt. This model assumes that final output (Y) is produced using labor and a continuum of different intermediate goods according to the production function:

$$Y = AL^{1-\alpha} \int_0^1 B(i)x^{\alpha}(i)\mathrm{d}i$$

where each $B(i)$ is a quality parameter indicating the productivity of intermediate good i. Each intermediate good is assumed to be produced according to the constant returns production function $x(i) = K(i)/B(i)$, where $K(i)$ is the amount of capital used to produce good i. Maximizing Y subject to $\int_0^1 AB(i)x^{\alpha}(i)\mathrm{d}i = K$ yields the optimality conditions for each $x(i)$ as

$$x(i) = x = K/B$$

where B is average quality and K is aggregate tangible capital. Hence the production function can be written as

$$Y = F(K, BL) = AK^{\alpha}(BL)^{1-\alpha}$$

Let $B^{\max}$ denote the maximum of all existing $B(i)$'s which may be called the leading edge technology. Each time an innovation occurs in a sector i it creates a new generation of intermediate good i with a quality parameter equal to the current value of $B^{\max}$.

Suppose that the economy-wide frequency of innovations is proportional to the amount of R&D: βn, where β is a positive parameter indicating the Poisson arrival rate of innovations. The leading edge technology grows over time as a result of the gradual accumulation of knowledge from innovations. Specifically, the exponential rate at which $B^{\max}$ grows is proportional to the frequency of innovations, i.e.,

$$\frac{\mathrm{d}B^{\max}}{\mathrm{d}t} = \sigma\beta n B^{\max}$$

where σ is a parameter at which the flow of innovations pushes out the economy's technology frontier. Note that this model displays unlimited growth since growth can be written as

$$g^* = (1/a)(\sigma\beta - \rho)$$

showing that the common rate of return $\sigma\beta$ to physical (K) and intellectual capital (B) does not diminish as more and more capital is accumulated.

Finally we note that Schumpeter's emphasis on knowledge-based growth is greatly underestimated by standard measures of R&D activity and resources used in the educational sector. Aghion and Howitt have pointed out that here we have a "knowledge investment problem," i.e., the output of knowledge resulting from formal and informal R&D activities in universities and public research institutions is typically not measured at all, because it does not result in an immediate commodity with a market price. Along with the investment problem there is the quality improvement factor associated with knowledge creation. For instance Gordon (1990) has estimated that correcting properly for quality improvements in capital goods alone would at least double the growth rate of aggregate real investment in the USA over the period 1947–1983.

5.4 Industry Applications

Innovation as the basis of industrial change has been central in Schumpeter's work in various ways. He placed innovation in the evolution of industries and within the process of economic transformation. This tradition has continued over the years and the recent changes in modern technology have provided ample support to Schumpeterian conjectures. The progress in the analysis of innovation and the evolution of industries has been on several fronts. Malerba (1985) has identified four major fronts as follows: First, major differences in innovation have been noticed across different industries. The effects of innovation differ in terms of different knowledge, competences, and learning processes. There is now enormous evidence on the contributions of universities, public research organizations, and financial organizations like venture capital. As Malerba (1985) has shown, the Schumpeterian patterns of innovation emphasize specific dynamics of innovation, firm entry, and growth which are different from one industry to another or from one sector to another. Secondly, great progress has been achieved in econometric work relating to innovation and its impact. All this work has identified robust interindustry differences in concentration, firm age distribution, and the characteristics of innovation. Thirdly, some innovation models formalized industrial life cycles, analyzing together product and process innovations, rate and type of entrants, market concentration, and market niche. Finally, progress has been made at a more macrolevel by linking innovation and industry evolution to structural change and the changing sectoral composition of the economy.

A. Information Technology Sector

In recent times competition has been most intense in the high-tech industries such as semiconductors, microelectronics, personal computers, and telecommunication. Norsworthy and Jang (1992) estimated the productivity growth in the US computer industry comprising mainframe, mini-, and microcomputers over three subperiods 1959–1967, 1967–1975, and 1975–1981. In a simplified form they estimated the translog cost function as

$$\ln \mathrm{TC} = b_0 + b_1 \ln y(t) + b_2 T$$

where $y(t)$ is output and T is a time trend variable used as a proxy variable for the state of technology. The inverse of the parameter b_1 can be interpreted as the degree of returns to scale. A negative value of b_2 indicating a pure shift of the cost function is a measure of Hicks-neutral technical progress. The regression estimate for b_2 turns out to be (−0.037) for the whole period 1959–1981, whereas $\hat{b}_1 = 0.348$ and both estimates are statistically significant at the 10 % level. This suggests that the long-run average rate of technical progress is 3.69 % per year in the US computer industry over the years 1959–1981 and the average degree of returns to scale is $1/\hat{b}_1 = 2.876$. More recent estimated for 2000–2010 have upheld this trend. Jorgenson et al. (2000) noted two significant impacts of the growth of computer power on the overall US economy. First, as the production of computers improves and becomes more efficient, more computing efficiency is being generated from the same inputs through learning by doing. This has steadily increased the overall industry growth. Secondly, the computer-using industries are now using skilled labor working with better computer equipment and better software, thus increasing overall labor productivity. Thus they estimated that the average industry productivity growth has achieved a rate of 2 % per year over the period 1958–1996 for electronic equipment, which includes semiconductors and communications equipment. Higher productivity growth led to falling unit costs and prices. Falling prices have led to increased demand due to high price elasticity. For instance, average computer prices declined by 18 percent per year from 1960 to 1995 and by 27.6 % per year over 1995–1998. This trend has continued over the current period 2000–2010.

The technological developments in the US computer and communications industry since 1981 have followed several distinct patterns in the new information age. One is the important role of R&D investments in both human capital and shared investment network. The theory predicts that technical uncertainties in R&D investment outcomes can be hedged considerably by pursuing multiple conceptual approaches in parallel. This reduces the expected time to successful project completion. Also successive R&D costs have been reduced consistently through networking and cost sharing among research institutions. The second important trend is the pattern of product cycle changes in recent years. Some researchers have argued that in the computer industry today the survival of the fittest depends on the need for a company to bring on line a continuous stream of new products. Each product goes through a typical life cycle of R&D, market introduction, maturation, and the eventual obsolescence. One trend is prominent. This is the trend of widening demand due to economies of scale in demand. Thus many early niche markets in the computer industry such as the markets for laptops, palm PCs, and smartphones have now grown into mass markets. As the manufacturers like Apple or Samsung scramble to shorten the time to market, the R&D and commercialization phase of each cycle becomes shorter. As the new products gradually penetrate an ever-expanding market, the upswing and maturation phases become longer and longer thus generating accelerated growth. Finally, in the new information age, the workplace has undergone dramatic changes in the new knowledge economy. The evolution from an economy based on traditional manufacturing to one base on knowledge and services has contributed to making old rigid managerial hierarchies less relevant. In 1972 manufacturing companies accounted for 70 % of the one hundred largest employers in the USA, UK, France, Germany, and Japan. By 2002 manufacturing's share had dropped to 30 %. This trend is continuing. Nations around the world face an acute shortage of skilled workers, even during recessionary times, because of the growing cap between the supply of workers educated with computer knowledge and the demand for their services. Many countries are also failing to create a long-term pipeline of workers having the skills required to participate in the knowledge economy. For example, China will need 75,000 business leaders by 2017, but only an estimated 3,000–5,000 were available in 2007. The US Department of Education estimates that 60 % of all new jobs in the early twenty-first century will require skills that only 20 % of the current workforce possesses. The persistence of these educational gaps imposes on the US economy the economic equivalent of a permanent national recession, states a 2009 McKinsey Report.

In the computer industry today no company stays with a single cycle due to a rapidly changing technology pattern. The managerial challenge is to bring an optimal stream of new vintages of its products and as each company does so, the process of technological evolution continues forward in the industry. Also there is a rapid convergence of technologies in the PC and communications field through iPad, iPhone, and smartphones.

We now discuss here some of the stylized facts about the computer industry and its recent upsurge. The speed of technical change, growth in demand, and heterogeneity of the firms are some of these important characteristics.

The impact of technical change can be modeled in terms of the change in market share s_i of firm i, which may be used as a proxy for entry or exit. The entry dynamics may take the form

$$\dot{s}_i = \lambda s_i(\bar{c} - c_i)$$

where dot denotes the time derivative, c_i is the minimal average cost of firm group i, and $\bar{c}$ is the industry average cost. Here λ determines the speed at which firm's market penetration adjusts to differences between c_i and $\bar{c}$. One may view $c_i = c(I_i) = -hI_i(t)$ as a function of innovations in R&D denoted by $I_i(t)$ with its parameter $(-h)$ reflecting its own cost-reducing aspect. The source of potential dominance of a firm is measured here by the gap $(\bar{c} - c_i)$, where $c_i < \bar{c}$. One of the findings of the entry–exit dynamics studied empirically by Lansburg and Mayes (1996) is that a growing competitive industry involves not just the expansion of existing firms but also new entrants who challenge the incumbents often with new innovations and thus the process of creative destruction of old processes or products begins.

We now consider a class of Pareto-type models to characterize the sources of productivity growth of firms and industries. These models are nonparametric in the sense that no specific forms of cost or production are assumed here. These models widely applied in management science and operations research are also called "data envelopment analysis" (DEA) since they involve the estimation of the convex hull of the input–output data. The impact of innovations as R&D and knowledge capital is analyzed here in terms of two types of models. One emphasizes the unit cost-reducing aspect, which helps lower the price and hence increases demand. Secondly, the impact on output growth through the increase in R&D inputs is formalized through a growth efficiency model. Here a distribution is drawn between *level* and *growth* efficiency, where the former specifies a static production frontier and the latter a dynamic frontier. Finally, the overall cost efficiency is decomposed into technical (TE) or production efficiency and allocative efficiency (AE).

Denote unit cost by $c_j = C_j/y_j$, where total cost C_j excludes R&D cost denoted by r_j for firm $j = 1, 2, \ldots, n$. Then we set up the nonparametric Pareto efficiency model as

$$\begin{aligned}
&\min \theta \text{ subject to} \\
&\sum_{j=1}^{n} c_j \lambda_j \leq \theta c_h \\
&\sum_j r_j \lambda_j \leq r_h; \ \sum_j y_j \lambda_j \geq y_h \\
&\sum_j \lambda_j = 1; \lambda_j \geq 0; \theta \geq 0 \\
&j = 1, 2, \ldots, n
\end{aligned}$$

On using slack variables $\alpha, \beta_0, \beta_1, \beta_2$ and solving the LP model above we obtain for an efficient firm h, with $\theta^* = 1$ and all slack variables zero, the following average cost frontier:

$$c_h^* = \hat{\beta}_0^* - \hat{\beta}_2^* r_h + \hat{\alpha}^* y_h$$

where $\hat{\beta}_0 = \beta_0/\beta_2^*$, $\hat{\beta}2 = \beta_2/\beta_1^*$ and $\hat{\alpha} = \alpha/\beta_1^*$. If we replace r_h by cumulative knowledge capital R_h as in Arrow's learning-by-doing model, then the AC frontier becomes

$$c_h^* = \hat{\beta}_0^* - \hat{\beta}_2^* R_h + \hat{\alpha}^* y_h$$

This shows that any increase in R_h reduces the unit cost of the efficient firm h, for which $\theta^* = 1$ holds. Clearly if $0 < \theta^* < 1$ in the above LP, then firm h is not Pareto efficient, since other firms or a convex combination of them have lower average costs. Thus the innovating firm which follows the cost frontier reduces unit costs in the long run and gains market share.

To consider the dynamic impact of R&D investments and innovation in different inputs we use a growth efficiency model in DEA framework, one based on a dynamic production frontier and the other on a dynamic cost frontier. We use Standard and Poor's Compustat database with SIC codes 3570/3571 over the period 1985–2000 covering 40 computer firms. The dynamic production frontier model uses a nonradial efficiency score $\theta_i(t)$ specific to input i as follows:

$$\min \sum_{i=1}^{m} \theta_i(t)$$

$$\text{s.t.} \sum_{j=1}^{n} \tilde{x}_{ij}(t)\lambda_j(t) \leq \theta_i(t)\tilde{x}_{ih}(t)$$

$$\sum_{j=1}^{n} \tilde{y}_j(t)\lambda_j(t) \geq \tilde{y}_h(t)$$

$$\sum_{j} \lambda_j(t) = 1,\ j \in I_n = (1, 2, \ldots, n)$$

$$i \in I_m = (1, 2, \ldots, m),\ t = 1, 2, \ldots, T$$

Here $\tilde{x}_{ij}(t)$ and $\tilde{y}_j(t)$ denote percentage growth rates of m inputs and n output. If firm j is efficient, it would follow the dynamic production frontier

$$\Delta y_j(t)/y_j(t) = \beta_0^* + \sum_{i=1}^{m} \beta_i^*(\Delta x_{ij}/x_{ij})$$

where β_0^* is free in sign for the equality constraint $\sum \lambda_j(t) = 1$ and the other multipliers β_i^* are nonnegative. Clearly the scale S is here measured by $S = \sum_i \beta_i^*$ and β_0^* measures Solow-type technical progress if it is positive. Thus by using a 4-year moving average, say one could obtain a long-run change in scale $S(\tau)$. Thus if $\beta_0^*(4) > \beta_0^*(3) > \beta_0^*(2) > \beta_0^*(1) > 0$, then the technology due to innovation is improving over time. Likewise $S(4) > S(3) > S(2) > S(1) \geq 1$.

A cost-oriented version of the of the above DEA model may be specified by the LP model:

$$\min \phi(t)$$

$$\text{s.t.} \sum_{j=1}^{n} \tilde{C}_j(t)\mu_j(t) \geq \phi(t)\tilde{C}_h(t)$$

$$\sum_{j=1}^{n} \tilde{y}_j(t)\mu_j(t) \geq \tilde{y}_h(t)$$

$$\sum_{j=1}^{n} \tilde{y}_j^2(t)\mu_j(t) = \tilde{y}_h^2(t)$$

$$\sum \mu_j = 1, \mu_j \geq 0, j \in I_n$$

Here we have used a radial efficiency score $\phi(t)$, instead of input-specific scores, and the growth of observed total costs and total output is denoted by $\tilde{C}_j(t) = \Delta C_j(t)/C_j(t)$ and $\tilde{y}_j(t) = \Delta y_j(t)/y_j(t)$. A quadratic output constraint is added here an equality constraint, so that the associated Lagrange multiplier γ^* can be unrestricted in sign. If the firm j is efficient, then its dynamic cost frontier can be written as

$$\tilde{C}_j(t) = \gamma_0^* + \gamma_1^* \tilde{y}_j(t) + \gamma_2^* \tilde{y}_j^2(t)$$

If one excludes R&D costs from total cost C_j and denotes it by $R_j(t)$, then the dynamic cost frontier becomes

$$\Delta C_j(t)/C_j(t) = \beta_0^* + \beta_1^*(\Delta y_j(t)/y_j(t)) - \beta_2^*(\Delta R_j(t)/R_j(t))$$

Here β_1^*, β_2^* are nonnegative and β_0^* is unrestricted in sign. Two comments are in order. First, the nonradial efficiency scores $\theta_i(t)$ may represent input-specific innovations, whereas the radial measure ϕ^* may denote the impact of all knowledge capital. Second, the quadratic cost frontier may be written as

$$\tilde{C}_j(t)/\tilde{y}_j(t) = \gamma_0^*/\tilde{y}_j(t) + \gamma_1^* + \gamma_2^* \tilde{y}_j(t)$$

Then the efficient firm j can choose an optimal growth rate of output by minimizing $\tilde{C}_j(t)/\tilde{y}_j(t)$ as

$$\tilde{y}_j = (\gamma_2^*/\gamma_0^*)^{1/2}$$

The empirical estimates for the computer industry are presented in several tables as follows: Table 5.1 presents the nonradial efficiency scores of the production frontier model, where all inputs are grouped into three inputs as R&D (x_1), net profit and capital expenditure (x_2), and cost of goods sold excluding R&D spending (x_3). The importance of R&D inputs is clearly revealed by its high efficiency scores.

Table 5.1 Nonradial average efficiency scores

	1985–1989			1990–1994			1994–2002		
Company	θ_1^*	θ_2^*	θ_3^*	θ_1^*	θ_2^*	θ_3^*	θ_1^*	θ_2^*	θ_3^*
1. Dell	0.61	0.44	0.47	1.0	1.0	1.0	1.0	1.0	1.0
2. Compaq	0.40	0.54	0.50	1.0	1.0	1.0	0.33	0.60	0.75
3. HP	0.49	1.0	0.47	0.55	0.80	1.0	1.0	1.0	1.0
4. Sun	1.0	1.0	1.0	1.0	1.0	1.0	0.42	0.24	0.67
5. Toshiba	0.49	0.62	0.72	1.0	1.0	1.0	1.0	1.0	1.0
6. Silicon Graphics	1.0	1.0	1.0	1.0	1.0	1.0	0.25	0.25	0.38
7. Sequent	1.0	1.0	1.0	1.0	1.0	1.0	0.50	0.54	0.48
8. Hitachi	0.40	0.68	0.65	1.0	1.0	1.0	0.94	0.84	1.0
9. Apple	0.52	0.69	0.64	0.51	0.44	0.76	1.0	1.0	1.0
10. Data General	1.0	1.0	1.0	1.0	1.0	1.0	1.0	0.48	0.54

Table 5.2 Impact of R&D on growth efficiency based on the cost-oriented model

	1985–1989		1990–1994		1994–2002	
Company	θ^*	β_2^*	θ^*	β_2^*	θ^*	β_2^*
Dell	1.0	2.71	1.0	1.15	0.75	0.08
Compaq	0.97	0.03	1.0	0.0	0.95	0.0
HP	1.0	1.89	0.93	0.10	0.88	0.0
Sun	1.0	0.0	1.0	0.13	0.97	1.79
Toshiba	0.93	1.56	1.0	0.13	0.97	1.79
Silicon Graphics	0.99	0.02	0.95	1.41	0.87	0.0
Sequent	0.72	0.80	0.92	0.0	0.84	0.0
Hitachi	0.88	0.07	0.98	0.21	0.55	0.0
Apple	1.0	1.21	0.87	0.92	0.68	0.0
Data General	0.90	0.92	0.62	0.54	0.81	0.65

Companies which have experienced substantial growth in sales have also spent heavily on R&D.

Note that the R&D spending includes here not only software development but also all types of marketing and network expenditures. The companies which are leaders in growth efficiency also show a very high elasticity of output with respect to R&D spending Table 5.2.

Now we consider a regression estimate for the impact of R&D input. The results are as follows:

$$\begin{aligned} y &= 70.8^* + 3.62^* x_1 + 0.29^{**} x_2 + 1.17^* x_3 \\ R^2 &= 0.981 \\ \Delta y &= -6.41 + 2.65^{**} \Delta x_1 + 1.05^{**} \Delta x_2 + 1.17^* \Delta x_3 \\ R^2 &= 0.994 \end{aligned}$$

Table 5.3 Estimates of demand growth

	$\hat{\beta}$	t-value	adj. R^2
Dell	1.49	41.1	0.994
Compaq	1.28	28.7	0.988
HP	1.12	26.7	0.986
Sun	1.11	31.8	0.990
Toshiba	1.04	9.5	0.899
Silicon graphics	0.99	9.5	0.899
Sequent	0.99	9.5	0.897
Hitachi	0.72	4.61	0.669
Apple	0.70	4.43	0.650
Data general	0.72	10.21	0.681

where one or two asterisks denote significant t-values at the 5 % and 1 %, respectively. When the regressions are run separately for the DEA efficient and nonefficient companies, the coefficient for the R&D input is about 12 % higher for the efficient firms.

Another important stylized fact in the computer industry is the upsurge in demand. It is one of the fastest growing sectors in the US economy over the 16-year period 1985–2000. The average sales growth of the companies in the SIC codes 3570 and 3571 is about 12.8 % per year on the average for the period 1985–1994 and slightly higher (13.1 %) for 1995–2000. Two aspects of demand upsurge are to be noted. One is the increase in demand volume due to globalization of trade. The trade expansion has been spearheaded by the rapid advance in software development by the subsidiaries of leading US companies in Asian countries like Taiwan, South Korea, China, Singapore, and India. The second aspect of demand growth is due to the significant economies of scale in demand. The elasticity of demand with respect to total industrial output has exceeded 2.91 over the whole period 1985–2000, whereas the income elasticity of demand has been about 1.92. The market size effect has also been very significant. The price elasticity of demand rose from −1.74 to −2.15 from 1985–1992 to 1993–2000. A rapidly growing market thus reflects high R&D spillover effects. The regression estimates for the Compustat data over 1985–2000 based on the equation

$$y_t = \alpha + \beta y_{t-1} + e_t$$

are given in Table 5.3. On the average for the industry as a whole the annual rate of demand growth has increased from 8 to 15 % . This has helped the leading firms increase their cost efficiency through R&D investments.

B. Information Technology Sector

The pharmaceutical industry provides a unique example of the challenge to the competitive model and its concept of efficiency. Which market structure induces the greatest incentives for investment in R&D and the related innovation? Do firms invest more in fragmented industries, where each firm is of small size and product market competition is very intense, or rather in industries where a few firms command significant market power? Schumpeter argued that perfect competition is not only less suitable for large R&D innovations but inferior and has no title of being set up as a model of ideal efficiency. The competitive paradigm ignores two strategic aspects of innovation: one is the need for large capital investment for several years as in the pharmaceutical industry. Secondly, a large fraction of R&D investments are self-financed and this indirectly explains why large firms may be better positioned to perform R&D. Also the economies of scale and economies of scope resulting from innovative R&D expenditures allow the large firms to spread the risk from large R&D investment more easily. This is more true for the pharmaceutical industry. From a Schumpeterian point of view the optimal market structure is not likely to be perfect competition but rather a form of dynamic competition that involves some degree of monopoly power. Many policymakers and legal experts subscribe to the view that perfect competition implies efficient resource allocation in a static sense, but that optimality breaks down when one takes dynamic efficiency into consideration.

In this connection it is important to note a significant trend of merger waves in the pharmaceutical industry that started in the late 1990s. The growth of managed care in the USA and the tightening of government health-care budget in other nations forced manufacturers in the pharmaceutical industry to lower prices on many drugs. In response to these pressures this industry underwent a remarkable wave of consolidation with the total value of merge and acquisition activity exceeding $500 billion. As a result the combined market shares of the ten largest firms in the USA have grown from 20 % to more than 50 %. Using almost any yardstick, we can view Glaxo's 2000 acquisition of SmithKline Beecham and Pfizer's acquisition of Pharmacia as among the largest in business history. The most important reason for this merge wave is to exploit economies of scale in R&D.

Azoulay and Henderson (2001) have noted that instead of vertical integration through mergers, the major drug houses have chosen tapered integration for drug development. Clinical trials are the most time-consuming and costly step in the pharmaceutical vertical chain. Large-scale double-blind research trials can take 10 to 15 years and cost over $100 million. Drug makers must therefore manage clinical trials effectively so as to be successful. The authors argue that firms tend to outsource "data-intensive projects" while keeping "knowledge-intensive projects" in-house. The underlying idea is that drug makers must develop research protocols in close concert with their research scientists.

Azoulay and Henderson found that research productivity of the pharmaceutical firms measured as the number of patents obtained per research dollar invested depends on three classes of factors: the composition of a firm's research portfolio, firm-specific scientific and medical know-how, and a firm's distinctive capabilities. Thus they found that investments in cardiovascular drug discovery were consistently more productive and profitable than investments in cancer research. The large size effect has helped to exploit significant economies of scale and the whole industry structure has moved in the direction that total industry profits are higher. This feature is often called "the efficiency effect." The learning curve also provides an additional

Table 5.4 Estimates of net sales and production costs for five pharmaceutical companies

	Net Sales				Production Costs			
	$\hat{a}$	$\hat{b}$	$\bar{R}^2$	DW	$\hat{a}$	$\hat{b}$	$\bar{R}^2$	DW
Abbott	317.7**	1.04**	0.99	1.05	27.7	1.07	0.99	2.47
B & L	175.2	0.915*	0.90	2.48	92.1	0.87**	0.80	2.48
Merck	−448.2	1.22**	0.99	2.67	−65.7	1.27**	0.98	1.9
Pfizer	−2649.5	1.55**	0.26	1.14	28.9	0.91	0.04	0.98
Pharmacia	−2579.6	1.41**	0.26	0.75	1954.3*	0.53**	0.26	1.3

source of persistent dominance by the market leader. By selling more the leader lowers its costs faster, which in turn makes it more competitive. This in turn makes it sell more and so on (Table 5.4).

We now analyze empirically the efficiency structure of firms in the pharmaceutical industry by using the Pareto efficiency models or DEA. The pharmaceutical companies have grown very fast over the last decade, as breakthroughs in medical research have led to the developments of new medicines and new procedures. An overview of demand growth and the production costs measured by cost of goods sold for five selected companies are reported here over the period 1982–2000 by using the following autoregressive equation:

$$y_t = a + by_{t-1}$$

where y_t may denote either total sales or total cost of goods sold. The five companies are Abbott Lab, Bausch and Lomb, Merck, Pfizer, and Pharmacia. Here one and two asterisks denote significant t-values at 5 and 1 %, respectively, and $\bar{R}^2$ denotes adjusted R^2. Note that sales growth rate as measured by $\hat{\beta} = \hat{b} - 1$ has been the highest for Pfizer (0.553) followed by Pharmacia (0.406), Merck (0.220), and Abbott (0.042).

We consider a larger set of companies for assessing the impact of R&D investments in research and innovations over the 19 years 1981–2000. A set of 17 companies out of a larger set of 45 is selected from the Compustat database available from Standard and Poor. This selection is based on considerations of continuous availability of data on R&D expense and its share of total costs. These selected companies comprise such well-known firms as Merck, Eli Lily, Pfizer, Bausch and Lomb, Johnson & Johnson, GlaxoSmithKline, Schering-Plough, and Genetech. Three types of estimates are calculated for the selected companies. Table 5.5 reports the estimates of efficiency coefficients (θ^*) for the total cost (TC) and average cost frontier (AC) found from the DEA LP model. Table 5.6 provides the regression estimates of the cost frontier for the whole period 1981–2000 in the form

$$\text{TC}_j = a + by_j + cR_j + d\theta_j$$

Finally, Table 5.7 reports the estimates of the market share model, which predicts that the efficient firms would increase their market share when the industry average cost rises due to the inefficient firms failing to reduce unit costs. For the whole

Table 5.5 Efficiency coefficients (θ^*) for total cost (TC) and average cost (AC) frontier in selected pharmaceutical industry

	1981		1990		2000	
	TC	AC	TC	AC	TC	AC
Abbott Lab	0.829	0.831	0.871	0.885	0.807	0.831
Bausch&Lomb	0.877	1.0	0.768	1.0	0.737	0.739
Bristol Myers	0.832	0.861	0.971	1.0	0.939	0.982
Genentech	0.264	0.273	0.549	0.559	0.545	0.556
Glaxosmith	0.493	0.514	0.787	0.818	0.847	0.964
Johnson&Johnson	0.958	1.0	1.0	1.0	0.938	1.0
Eli Lily	0.772	0.886	0.781	0.811	0.840	0.903
Merck	0.710	0.848	0.983	1.0	1.0	1.0
Pharma	0.548	0.196	1.0	0.199	0.680	0.442
Pfizer	0.764	0.822	0.838	0.822	0.841	1.0
Schering	0.703	0.709	0.796	0.808	0.837	0.872

Table 5.6 Cost frontier estimates of selected pharmaceutical companies $TC_j = a + by_j + cR_j + d\theta_j$

Firm	a	b	c	d	R^2
ABT	1355**	1.301**	−295.3*	−0.046	0.999
B & L	−250.3	1.608**	–	−1375.4*	0.996
GSK	1915.1**	1.335**	−10929.1	−1595.7*	0.993
PHA	−16807*	2.224*	–	−1.517*	0.985

Table 5.7 Changes in market share (s) for selected companies in pharmaceutical industry over 1981–2000 $\Delta s = b_0 + b_1(\bar{c} - c)$

Firm	b_0	b_1	R^2
ABT	1.07**	2.267**	0.385
B & L	1.095**	0.691^a	0.296
BM	1.022**	1.577*	0.224
GSK	1.082**	0.727	0.027
PF	1.209**	12.271^{**a}	0.492
PHA	1.121**	3.947*	0.198

Note: The superscript a denotes that the quadratic term $(\bar{c} - c)^2$ has a significant positive coefficient.

industry over the period 1981–2000 this relationship is tested by the following regression, where $\bar{c}$ is the industry average unit costs including both efficient and inefficient firms:

$$\bar{c} = 3953.2^{**} + 5.79^{**}\sigma^2,\ R^2 = 0.912$$

$$(t = 10.02)$$

$$\Delta\bar{c} = 0.145^{**} + 0.005^{**}\sigma^2 - 0.014^9\bar{c},\ R^2 = 0.721$$

$$(t = 2.31)\quad (t = 3.45)$$

and σ^2 denotes variance of unit costs. Note that the coefficient of variance σ^2 is in units of E-05. Thus higher variance tends to increase average industry costs and thus allows efficient firms to increase their market share. Note that the impact of variance is highly significant statistically. Clearly the "churning effect" is found to be important. The efficiency effect we discussed before is applicable here.

Several points emerge from the estimated results. First of all, the number of firms on the cost efficiency frontier is about one-third. The following estimates report the number of efficient firms out of 17 which are efficient in terms of TC, AC, and R&D inputs. Growth of R&D inputs is very significant in this industry. It reduces

	TC	AC	R&D
1981	3 (18 %)	6 (35 %)	6 (35 %)
1990	5 (29 %)	6 (35 %)	6 (35 %)
2000	3 (18 %)	5 (29 %)	5 (29 %)

unit costs and increases profit through higher demand due to lower prices and lower unit costs. Secondly, the market share model shows that the efficient firms with $c < \bar{c}$ increase their market share over time and the efficiency of R&D inputs is very significant here.

C. Innovation in Service Industries

Innovation in the service sector is difficult to evaluate because of the fuzzy nature of the product of the service sector. Griliches (1992) has argued that most of the problems of output measurement in the service sector are due to the following reasons:

1. The nature and content of the transaction (e.g., in the service sector provided by the physician, is it the procedure or the results of care?)
2. The nature of the user involvement in the definition of the service output
3. The quality change that is more difficult to detect in services and to account for in price structures

The hedonic pricing techniques have been frequently applied here to adjust price indexes so as to account for quality changes.

Gallouj and Savona (2008) have provided an excellent review of the impact of innovation in services. We discuss below their review in some detail. One of the seminal contributions on the nature of innovation related to firm growth is that of Abernathy and Utterback (1978) who developed a dynamic model linking the competitive strategy of the firm with the development of the production process and the pattern of innovation. As the market matures firms tend to implement process development through exploitation of new technological opportunities rather than market need. This brings about product differentiation through segmented process development. Finally, as markets become saturated, innovative firms adopt cost-minimizing strategies through a process of systematic development aimed at standardizing their production. Barras (1990) adopted this framework to develop a dynamic model of innovation to account for the specificity of the production processes in service industries. In his Reverse Product Cycle model Barras has argued that ICTs represent the "enabling technology" created elsewhere and adopted by the service sector. Three evolving stages are identified: incremental process innovation, radical process innovation, and finally product innovation. The first stage is characterized by a "supplier-dominated" stage where the adopting service industry develops an incremental process innovation aimed at increasing the efficiency of the production process. In the second stage, characterized by a higher point in the learning-by-doing curve, radical process innovations take place aimed at improving the quality of service provided. The third stage is "user dominated" in the sense that the expansion of technological opportunities allows the introduction of radical service innovations.

Drawing on a data set of more than 2,000 sample observations on significant innovations and innovating firms in the UK over the period 1945–1980 mainly relating to the sectors using the innovation, Pavitt (1984) has discussed four categories of innovation trajectories as follows: (1) US supplier dominated; (2) production intensive, e.g., scale intensive; (3) information intensive; and (4) science-based suppliers. Gallouj (2002) has used the Lancasterian characteristics-based approach to the definition of product to develop a richer set of innovation models applicable to the service sector. The following models provide some examples:

1. "Radical innovation" created a new set of vectors of competences and technical and service characteristics, e.g., twitter accounts in social networking.
2. "Improvement innovation" occurs when the set of vectors of characteristics remains unchanged by the quality value increases.
3. "Incremental innovation" occurs when a new characteristic is added, eliminated, or substituted, e.g., the successive windows are common examples.
4. "Ad hoc innovations" relate to knowledge intensity activities. Those typically result in a new solution to a client problem.
5. "Recombination innovation" may involve the creation of a new product as a combination of characteristics of one or more products. Alternatively it may create a new product by means of fragmentation.
6. "Formalization innovation" occurs when one or more characteristics of the product or service are standardized.

It is important to mention here some of the broad trends of innovation in the service sector. First, there has been a tremendous growth in the social networking, e.g., Facebook and Twitter accounts, and the worldwide demand has expanded. Second, there has been a consistent trend in miniaturization of technology and innovative information and this is likely to continue. Thirdly, in case of financial services, service innovations have multipliers in recent times repeating a radical process innovation. This has led to substantial improvement in information efficiency for the users and suppliers. Finally, the technological dimension of innovation in service industries emerged recently with the diffusion of ICTs mostly in business services. This is now called "the knowledge-intensive business services.' Here the assimilation and convergence of technological trajectories remain dominant.

D. Growth of the Telecom Industry

Schumpeter considered innovation as key to industrial change and evolution. It was central to his work in various ways and forms. He was much interested in innovation either as a process of creative destruction or as a process of creative accumulation. He placed innovation in the evolution of industries and within the process of economic transformation.

The telecom industry provides a unique example of an industry which has evolved through many innovative transformations. As a communications industry it has its dynamic impact on numerous industries today. The telecommunications industry has evolved rapidly in the last two decades, when the advancement of the wireless technology sharply eroded the dominance of the fixed-line providers. Sengupta and Fanchon (2009) have discussed in some detail the growth of this industry and its efficiency aspects. We follow their discussion in brief and discuss the economic aspects of efficiency and the market competition. Three interacting forces drive the transformation of the industry: (1) market structure and deregulation, (2) innovation in many forms, and (3) significant capital expenditures by the firms with scale economies and spillover effects. Laffont and Tirole (2000) have discussed in some detail the main interaction between these forces which affect entry and efficiency in this industry. The telecom industry worldwide has been transformed by progressive deregulation and rapid change in technology, which facilitated the creation of many competitive firms. For example, the number of private telecom operators has increased from under 10 % to nearly 45 % between 1980 and 2000. The trend toward privatization has generated profound changes in the use and diversification of resources and in the kind of services provided. New mobile telephones offer the ability to connect to the Internet, iPad, and TV, download music and movies, send pictures, and even deposit checks to the banks. Apple, Samsung, and AT&T for example have spearheaded iPhones that can perform multiple tasks like taking pictures and sending them anywhere, following the social networks like Facebook and Twitter, and send messages anywhere. To remain competitive, any traditional telecom firm offering fixed-line connection had to invest in new technology so as to address the increasing demand for Internet services.

Privatization had not only stimulated competition but also forced the whole industry to go through a massive reallocation of its assets. Most of the investment has been in capital rather than labor and the industry has experienced rapid growth in total factor productivity. The link between privatization and productivity of the telecom industry has been explored in some detail by Li and Xu (2004). Using empirical data from the International Telecom Union and the World Bank-Stanford telecom project they propose a general model to explain the effects of privatization and market competition on a number of outcome variables like employment, investment, price of services, and productivity. The explanatory variables include two groups: reform variables which capture the effects of privatization and the control variables which capture the economic environment. The estimated regression results indicate that full privatization of the industry has a large and statistically significant positive effect on investment. The intensity of competition also increases the level of investment substantially. Li and Xu have also found from their production function estimates that deregulatory reforms have a large positive and significant effect on total factor productivity (TFP) especially full privatization which increases TFP by 32.5 % and the level of competition which increases TFP by 13.1 %.

Two methods are commonly used to measure efficiency in the telecom industry: one is parametric and the other nonparametric. We use the nonparametric DEA model of Pareto efficiency to identify the successful technology leaders in the telecom industry. The following input-oriented model is employed:

$$\min \theta \text{ s.t.}$$

$$\sum_{j=1}^{N} x_{ij}\lambda_j \leq \theta x_{ih}, i = 1, 2, \ldots, m$$

$$\sum_{j} y_j\lambda_j \geq y_h; \sum \lambda_j = 1, \lambda_j \geq 0$$

Here m is the number of inputs x_{ij}, y_j is output of firm j, and λ_j denoted a nonnegative weight defining the convex combination of inputs of different firms. The Compustat data set for codes SIC 4812 and 4813 with 4274 observations are used for the years 1998–2007. This data set was reduced to 2733 observations after excluding small firms with a value of yearly sales less than $100,000. Five variables were originally considered inputs:

x_1 = research and development expenses

x_2 = capital expenditures

x_3 = net value of plant and equipment

x_4 = cost of goods sold

x_5 = advertising expenses

The value of sales was used as a measure of output (y). A stepwise regression model is used in the form

$$y_t = \beta_0 + \sum_{i=1}^{5} \beta_i x_{it} + \sum_{i=1}^{5} \tilde{\beta}_i x_{i,t-1} + e_t$$

where $x_{i,t-1}$ represents the first-order time lag of input x_i. We found that the advertising input (x_5) is not statistically significant at 5 % level for any of the regressions with and without time lags. Also, the cost of goods sold is subtracted from the value of sales to obtain a measure of net output $\tilde{y}$. Finally our DEa model uses the following input variables: $x_{1,t-1}$ for research and development expenses with a lag, x_2 for capital expenditures, and x_3 for net plant and equipment. For comparison the DEA model was run twice: once with net sales $\tilde{y}$ and once with gross sales y. The two formulations gave closely similar results.

In order to obtain a clear identification of the variables that contribute most to efficiency, we exclude marginally efficient firms from the set of inefficient firms. The set of inefficient firms is thus identified as those with the efficiency parameter θ less than 0.90. Of the 363 firms in the study it turns out that 190 firms are fully efficient with $\theta^* = 1$ and 149 are inefficient with $\theta^* < 0.90$. We then proceed to estimate the sensitivity of the efficiency parameter θ^* to changes in the inputs in the following way. For each of the 149 firms selected one of the three inputs is increased by 1 % while the other two inputs are unchanged. The inputs and output of the other 190 efficient firms are left unchanged. We then compute the elasticity $\epsilon_{\theta i}$ $(i = 1, 2, 3)$ of efficiency θ with respect to input x_i. Similarly we estimate the elasticity for the 190 efficient firms. The results are as follows: It is clear that for

	$\epsilon_{\theta 1}$	$\epsilon_{\theta 2}$	$\epsilon_{\theta 3}$
Inefficient firms	0.081	0.818	0.080
Efficient firms	0.009	0.167	0.023

both efficient and inefficient firms an increase in capital expenditure has a far greater impact on efficiency than an increase in R&D expenditures. Also there is a striking difference between the elasticity for efficient and inefficient firms. For efficient firms the elasticity estimates also confirm that R&D has the most effect on efficiency with 1 year lag.

However one should note that the success of R&D projects is influenced by cumulative experience. Firms who engage in R&D usually make a long-term commitment. Hence a better measure of the benefits of R&D is the cumulative expenditure on R&D over a period, i.e., the learning-by-doing effect. The DEA model here did not analyze this effect.

5.5 Innovation and Industry Evolution

Two aspects are important for relating innovation to industry evolution. One is the demand growth and the other networking. Standard economic analysis claims that demand provides the incentives to innovation during industry evolution. The size, growth, structure, and composition of demand differentiation and market segmentation affect innovation in various ways in different stages of the evolution of an industry. Consumer behavior plays a major role in affecting the presence of information asymmetries and imperfect information with respect to new markets, submarkets, etc.

The challenge for networks starts from the recognition that innovation and industry evolution are highly affected by the interaction of heterogeneous agents with different knowledge, competences, and specialization. The evolutionary theory of firm growth has stressed the point that networks emerge because agents are different, thus integrating complementarities in knowledge, capabilities, and specialization. The diffusion of spillover effects of R&D provides another basic incentive for the emergence of networks affecting innovation and industry evolution.

Chapter 6
Innovation Challenges

Recent trends in modern technology and market dynamics have posed significant challenges for innovation and its impact on the evolution of several modern industries. Thus, in telecommunications equipment and services sector, a convergence of different technologies has taken place. This convergence has been associated with the creation of a wide variety of different agents ranging from large equipment producers to new service firms. In pharmaceutical and biotechnology sector a wide variety of science and engineering fields are playing important roles in widening the research space. Universities, venture capital, and health-care systems play a major role in the innovative process. In software industries a differentiated knowledge base has created several distinctive product groups. The role of large computer suppliers in developing integrated hardware and software systems has been displaced by many specialized software companies innovating either in package software or in customized software. The role of the university and public research institutions has become more important in the open source domain in diffusion of innovation and open competition. All these changes have profound implications for management and market dynamics. We discuss here in some detail the following innovation challenges:

1. Managerial challenge
2. Market challenge and competition
3. Network challenge
4. Challenge of knowledge creation and diffusion.

6.1 Managerial Challenge

The greatest challenge to innovation arises perhaps through riskiness of an innovative investment project. Teece (1986) argued that a new firm's ability to prosper from its innovations depends on the presence of a "market for ideas," where the firm can sell its ideas for full value. He identified two elements of the

J. Sengupta, *Theory of Innovation: A New Paradigm of Growth*,
DOI 10.1007/978-3-319-02183-6_6,

commercialization environment that affect the market for ideas: (1) the technology is not easily expropriated by others, and (2) specialized assets such as manufacturing or marketing capabilities must be used in conjunction with the innovative product. Also industries have undergone a drastic transformation from large-scale material manufacturing to the design of new technologies exhibiting increasing returns (IR) to scale. These are mechanisms of positive feedback that act to reinforce that which gets success. They occur due to three reasons: (1) high fixed costs and very low variable costs, (2) network effects which imply that the value of a product increases as the number of users rises, and (3) high switching costs implying a lock-in for the early customers.

All these factors imply a high degree of riskiness of all innovative projects. The uncertainty in the R&D process, the timing of R&D investments, and the threat of potential entry pose significant managerial challenges. Besides following optimal rules of investment under conditions of uncertainty, diversification provides another suitable outlet for risk minimization. One motive for diversification may be to achieve economies of scope and also economies of scale.

D'Aveni (1994) argues that in industries ranging from electronics to airlines and computer software the sources of competitive advantage are being created and eroded at an increasingly rapid rate. He calls this phenomenon *hypercompetition* and argues that a firm's chief strategic goal should be to disrupt existing sources of advantage in its industry including its own and create new ones. This idea reminds us that in environments characterized by rapid technological developments, a firm that rests on its laurels seeking only to harvest existing sources of advantage can be quickly displaced by more innovative rivals.

The incentives to innovate are affected by several factors such as sunk cost, the replacement effect, and the efficiency effect. The *sunk cost effect* arises due to the asymmetry between a firm that has already made commitments to a particular technology or product and one that is planning such a commitment. For an incumbent firm the costs associated with already incurred investments are sunk and thus should be ignored when the firm considers switching to a new technology. This creates inertia for the established firm to stick to the new technology. Schumpeterian creative destruction implies that all such inertia should be rejected in favor of the creative new technology or new product. Sutton (1991) has developed an important concept of *endogenous sunk cost.* Early in an industry's life cycle many small firms compete on a level playing field. But some firms grow larger than others due to superior quality or better customer service. Smaller firms now have one or two options. They can try to establish their own brand capital or they can differentiate and fill niches not exploited by the leaders. In these markets the sunk costs that limit the number of large firms are not the fixed up-front cost of production but the sunk cost of establishing a brand name. This type of cost is called "endogenous sunk cost" by Sutton. These costs may act as potentially deterrent for other firms' entry.

The replacement effect which may force firms not to innovate was considered by Arrow (1962). He showed that assuming an equal innovative capabilities, an entrant would be willing to spend more than the monopolist incumbent to develop the innovation. The reason is as follows: a successful innovation by the new entrant leads to a monopoly; a successful innovation by the established firm also leads to a monopoly, but since it already has a monopoly its gain from innovation is less than that for the potential entrant. This replacement effect explains why an established firm would be less willing to innovate or develop new sources of advantage than a potential entrant.

The efficiency effect makes an incumbent monopolist's incentive to innovate stronger than that of a potential new entrant. The reason is that the entrant not only takes business from the monopolist due to duopoly but also tends to drive down prices. This efficiency effect may dominate the other two effects, when the monopolist incumbent's failure to develop the innovation means that new entrants almost certainly will. In this case a key benefit of innovation to the established firm is to stave off the deterioration of profits from additional competition.

The modern age of innovations has ushered in a dramatic change in the managerial style of the corporate world. As Benko and Anderson (2010) have powerfully argued, the world of work today is at an inflection point. The hierarchal corporate ladder is giving way to a multidimensional corporate lattice. We now discuss in brief some of the salient features of the managerial transition from the ladder to lattice structures. In the corporate world the lattice model has helped innovative firms organize and advance a company's existing innovative efforts into a comprehensive strategic response to the altered landscape. It recognized that managerial career and life are no longer separate spheres but are now interdependent. Three core areas are central to the lattice model: how careers are built, how company work gets done, and how participation in the organization is fostered. The lattice depicts managers' career paths as multidirectional with moves across and down as well as up. A most important implication of the changing innovative world of work is that different things motivate different people. The interplay of creative managers unique motivations is termed as "career–life fit." This circumstance creates a performance challenge for companies today: how to *engage* members of their managerial workforce when individuals value a career–life fit that is unique to them and how to keep them engaged as their careers and lives change over time.

Engagement represents the extent to which employees and managers go the extra mile to deliver extraordinary results for the company internally and to serve as brand ambassadors externally. Benko and Anderson have noted survey results of 13 thousand US workers across major industrial sectors which found that companies with highly engaged employees and managers to those with moderate to low engagement enjoy 13 % higher total returns to shareholders and 26 % higher revenue per employee.

Engagement is a strategic input every bit as important as innovative capital and skilled labor. Indeed it is the engine of stellar individual performance that drives the organizational performance of the highly successful innovative companies today. The authors found that companies with high engagement scores deliver better results than those with low scores as follows:

1. Earnings per share growth is 160 % higher.
2. Return on assets is 100 % higher.

3. Revenue growth is 150 % higher.
4. Profitability is 40 % higher.
5. Productivity is 78 % higher.

Lattice organizations tend to instill a twenty-first century attitude toward transparency. As technology created unprecedented access to information, it has become easier to see into companies and measure their productive efficiency. The authors have cited examples like NASA, Cisco, and Microsoft, which show that the lattice structure has flattened communication and information channels presenting organizational structures an opportunity for more innovative organization. Not only does the workforce's collective know-how generate ideas for measurable improvements, but it also creates an environment that is highly productive. It seems that the Schumpeterian concept of creative accumulation following creative destruction has completed the full cycle in total engagement.

Recently Raynor and Ahmed (2013) have studied thousands of companies facing challenges of innovative technologies in a number of industries such as semiconductors, medical devices, electrical wiring, clothing retail, pharmaceuticals, and retail department stores and found three rules of success in management:

Rule 1: Better before cheaper
Rule 2: Revenue before cost
Rule 3: No other rules

Quality improvement is the most important goal for successful innovating firm. This aspect is the most critical for all incremental innovations. Cost reduction comes later. The second most important characteristic of a successful firm is revenue growth. This emphasizes the need for advertisement and well-planned marketing strategies.

6.2 Marketing Challenge

Three aspects of market challenges to innovation are most important for the evolution of industries. One is innovation competition. When several firms are competing to develop the same product or process improvement, the firm that does so first can gain significant advantage. This first mover advantage provides one of the strongest incentives to innovative investment. Secondly, innovation by large firms seeks to increase their dominance in market share in the industry, but since this innovation frequently has large spillover effects, the threat of potential entry is always present. This creates another important source of market rivalry and competition. Finally, market demand played a key role with respect to the emergence of so-called disruptive technologies, as Christensen (1997) has documented in the case of hard disk drives, earthmoving equipment, retail stores, and electrical motor equipment. We discuss in brief these three aspects in the following sections.

Innovation competition involves *patent race*, which describes the race between firms to innovate first. The first firm to succeed the innovation project wins the patent race and obtains the exclusive patent rights to develop and market the new product or new process. The patent right may provide the successful innovator to enjoy quasi-monopoly profits for 15 to 20 years. Economic models of patent races emphasize the importance of uncertainty in the R&D process and potential entry. The new entrants have to assess the following factors:

1. How much does the investment increase its R&D productivity and thereby increase the chances of winning the patent race?
2. Would other competitors increase their R&D spendings also thereby decreasing the new entrant's chances of winning the patent race?
3. How many competitors are there? If there are diminishing returns to R&D, the several small R&D firms may be a bigger threat to successful innovation by the new entrant that is a single competitor. If there are increasing returns, then one large conducting extensive R&D may be a more formidable competitor.

Besanko et al. (2010) cite the famous case of patent racing and the invention of integrated circuits (IC). The race to develop the first IC had two key protagonists: Jack Kilby of Texas Instruments (TI) and Bob Noyce of Fairchild Semiconductor. Both Kilby and Noyce found ways to combine transistors, resistors, and capacitors in a single unit with essentially no wires. In 1959 both filed patents for the designs of their semiconductors. After a 10-year battle the courts upheld Noyce's patents. Both Fairchild and TI continued to refine their integrated circuits, while the court case lingered and the two agreed to share royalties from any use of either design. Today both Kilby and Noyce share credit for inventing the IC. Kilby went on to invent the handheld calculator for TI, while Noyce founded Intel.

This example illustrates some key ideas about patent racing. TI and Fairchild were not the only firms attempting to create ICs and they succeeded for different reasons. Path dependence was behind each firm's success. Secondly, both firms made relatively large investments in research talent. Lastly, both firms understood that it is easier to form partnerships and other sharing arrangements before becoming product market competitors, when antitrust laws may stand in the way.

The first mover advantage earned by the successful innovating firm may be largely due in many cases to large investments in R&D and knowledge capital. This paves the way for market dominance, where there always exists the threat of potential entry. Cellini and Lambertini (2009) have formulated a dynamic R&D model for process innovation where $q_i(t)$ are the outputs ($i = 1, 2$) and the market demand and unit cost functions are

$$p(t) = A - q_1(t) - q_2(t)$$
$$\dot{c}_i(t)/c_i(t) = \delta - k_i(t) - \beta k_j(t), i \neq j$$

where dot denotes time derivative, $k_i(t)$ is the R&D effort of firm i, and δ is the constant rate of depreciation. The positive parameter β ($0 < \beta < 1$) denotes the technological spillover that firm i receives from firm j. When each firm behaves independently, the cost of R&D effort is assumed to be of the form

$$C(k_i(t)) = b(k_i(t))^2, b > 0$$

Assuming the case of independent R&D ventures, each firm maximizes a discounted profit function

$$\max_{q_i(t)} \pi_i = \int_o^\infty \exp(-\rho t)[(\lambda - q_i(t) - q_j(t) - c_i)q_i(t) - b(k_i(t))^2]\mathrm{d}t$$

subject to the cost and demand constraints. On applying Pontryagin's maximum principle the optimal trajectories turn out to be

$$q_i^*(t) = (1/2)(A - q_i(t) - c_i(t))$$
$$k_i^*(t) = (1/2b)(-\lambda_{ii}(t)c_i(t) - \beta\lambda_{ij}(t)c_j(t))$$

where $\lambda_{ij}(t) = \mu_{ij}(t)e^{-\rho t}$ is the present value costate variable for the control variable $c_i(t)$. These two conditions describe the standard Cournot–Nash reaction functions. If we satisfy the conditions $\delta\rho \leq \frac{A^2(1+\beta)}{24b}$, then there is a saddle point equilibrium with steady-state levels:

$$\bar{c} = [2(1+\beta)]^{-1}[A^2(1+\beta)^2 - 24b\rho\delta]^{1/2}$$
$$\bar{k} = \delta(1+\beta)^{-1}$$

where $c_1(t) = c_2(t) = c(t)$ because of symmetry. Clearly $\partial\bar{k}/\partial\beta < 0$, which implies that an increase in the spillover effect β leads to a decrease in the level $\bar{k}$ of R&D. One way to restore the higher level of R&D is through public subsidies. As Spence (1984) has argued, the R&D output has the character of a public good. Hence it is socially optimal to supply the public good publicly or subsidize the private supplier, who should at least be rewarded for the benefits it confers on other firms.

In case the two firms form a cartel, the firms choose output levels noncooperatively while maximizing joint profits. Then steady-state levels are such that the level of $\bar{k}$ remains the same but unit cost $\bar{c}$ is lower. The extent of consumers' surplus in steady state however is much lower in case of cartel compared to the case of independent ventures.

Modern firms in this information technology-driven world today have several economic reasons to cooperate and combine R&D efforts. First of all, the technology of the new innovation processes is becoming increasingly complex and the fixed cost of developing the product or the process is becoming very large because this frequently generates dynamic returns to scale which result in entry barriers and therefore imperfect price competition. Secondly, there is increasingly the possibility that competitors may imitate the new technology. Thirdly, collusion and cooperation in the R&D phase may help the innovating firms to internalize a large portion of the spillover effects and thereby reduce unit costs which may lead to larger market shares. By now the governments in most advanced industrial countries have recognized this need. For example the European commission allowed in March 1985 a 13-year block exemption under Article 85(3) of the Treaty of Rome to all firms forming joint ventures or cartels in R&D. Thus antitrust laws would not apply to such cartels.

The market demand and its growth provides another challenge for industrial evolution. When the relations between demand and innovation are examined, one has to mention the whole literature on diffusion for competing technologies. As Malerba (1985) has shown, coinvention has followed the growth of market demand. Thus innovation by sellers and complementary investments and innovation by buyers (in terms of new products and services and investments in human capital) have played a catalytic role in the growth of innovative investments. Thus in software industries (open source software) communities of practices are major sources of innovation and change. They act as facilitators of innovation, because members who innovate are able to share their ideas with other members, assist them, and even obtain resources to develop their innovations.

6.3 Network Challenge

This challenge starts from the interaction of heterogeneous agents with different knowledge and specialization, who affect innovation and the industry evolution. Recent upsurge of social networks like Twitter, Facebook, and Linkedin has resulted in phenomenal growth in advertising and as a consequence the economies of scale in demand have played a very dynamic role. Cowan and Ozman (2004) have shown that in industries in which tacit knowledge is important and technological opportunities are high, regular industry structures generate higher knowledge growth, while in industries in which knowledge is codified and technological opportunities are lower, communication without any structure performs better. Networks of innovations are frequently driven by technological complementarities. Coevolution of innovations and technologies play the most important role here. In computers coevolutionary processes involving technology, minicomputers, and computer networks have differed greatly in their diffusion. In pharmaceuticals and biotechnology, the interactions between knowledge, technology, and country-specific factors have shaped the industry evolution.

6.4 Knowledge Creation and Diffusion

Economic studies of the evolutionary theories of the firm have shown that different sectors of the economy and different technologies vary greatly in terms of their knowledge base and learning processes related to innovation. In some sectors like communications and genetics science is the main force driving knowledge growth, while in others like software learning by doing and cumulativeness of advancements are the major forces. Knowledge has also different degrees of accessibility with major consequences on entry and concentration and may be more or less cumulative.

In high-tech industries today investments in R&D and knowledge capital have played a crucial role as engines of growth. Several dynamic features of R&D investment by firms are important for industry evolution. As measures of knowledge creation patent citations, R&D spending and research publications are often quoted. R&D spendings are the most important of these indicators. R&D expenditure not only generates new knowledge and information about new technical processes and products but also enhances the firm's ability to assimilate, exploit, and improve existing information and knowledge capital. Enhancing this ability to assimilate and improve existing knowledge affects the learning process within an industry that has cumulative impact on industry evolution. Cohen and Levinthal (1989) (C&L) have shown that one of the main reasons firms invested in R&D in the semiconductor industry is because it provides an in-house technical capability that could keep these firms on the leading edge of the latest technology. Secondly, R&D yields externalities in the sense that knowledge diffusion occurs between firms, which find new applications both locally and globally. Finally, the possibility of implicit or explicit collaboration in R&D networking and joint ventures increases the incentives to invest more. C&L have analyzed two facets of R&D investment: one is to reduce the incentive to investment because of the spillover effect and the other to encourage equilibrium investment at the industry level while other firms contribute to the overall knowledge capital. C&L empirically estimate by regression (OLS, GLS, and Tobit) models the effects of the knowledge inputs and other industry characteristics on unit R&D spendings (i.e., R&D intensity) of business units. The model is of the form

$$z_i = M_i + a_i \left(\theta \sum_{j \neq i} M_j + T \right), 0 \leq a_i \leq 1$$

where z_i denotes the firm's stock of technological and research knowledge, M_i the firm's R&D investment, a_i is the fraction of intra-industry knowledge that the firm is able to exploit, and θ is the degree of intra-industry spillover of research knowledge. It is assumed that $a_i = a_i(M_i, \beta)$ depends on both M_i and β, where β is a composite variable reflecting the characteristics of outside knowledge such as complexity and ease of transferability. Here θ denotes the degree to which the research effort of one firm may spillover to a pool of knowledge potentially available

Table 6.1 Effects of knowledge and other explanatory variables on R&D intensity

	OLS	GLS	Tobit
1. Technological opportunity			
(a) Appropriability $(1-\theta)$	0.396*	0.360**	0.260
(b) User tech	0.387**	0.409**	0.510**
(c) Univ tech	0.346**	0.245**	0.321*
(d) Govt tech	0.252*	0.170*	0.200*
2. Basic science research			
(a) Biology	0.176	0.042	0.159
(b) Chemistry	0.195**	0.095	0.149
(c) Physics	0.189	0.037	0.156
3. Applied science research			
(a) Computers	0.336**	0.157	0.446**
(b) Material science	−0.005	−0.028	0.231*
4. New plant	0.055**	0.041**	0.042**
5. Elasticity of			
(a) Price	−0.180**	−0.071	−0.147*
(b) Income	1.062**	0.638**	1.145**

Note: One and two asterisks denote t-values significant at 5 % and 1 % respectively

to all other firms also. Thus $\theta = 1$ implies that all the benefits of one firm's research accrue to the research pool available to all others, whereas $\theta = 0$ implies no spillover.

The sample data includes R&D performing business units consisting of 1302 units representing 297 firms in 151 lines of business in US manufacturing over the period 1975–1977. Data are from FTC's Line of Business Program and some survey data collected by C&L. Respondents of survey data were asked to rate on a 7-point scale the effectiveness of different methods used by firms to protect the competitive advantages of new products and new processes. For a given line of business appropriability is then defined as the maximum score. Thus if appropriability increases, the spillover level declines and hence R&D intensity increases. Some broad conclusions emerge from the estimated reported in Table 6.1. First of all, the estimated results across all the three estimation methods have the hypothesis that the effects on R&D intensity of the basic and applied science are equal. This implies that the role of learning differs significantly across field. Secondly, increasing technological opportunity through the less targeted basic sciences evokes more R&D spending than does the applied science. Finally, the OLD (ordinary least squares) and GLS (generalized least squares) of the coefficient of appropriability are positive and significant. This implies that spillovers have a net negative effect on R&D intensity. Also both user technological opportunity and university technological opportunity have strongly significant impact on R&D intensity.

Now consider knowledge diffusion and the role of entrepreneurship that turns knowledge into innovative products. Block et al. (2012) have used empirical evidence using European country-level and pooled OLS fixed and random effects regressions to show that a high rate of entrepreneurship increases the chances that knowledge will become new-to-the-market innovation. These findings highlight the importance of Schumpeterian entrepreneurship in the process of commercialization of knowledge. We discuss briefly their contribution to entrepreneurship and innovation policy.

The knowledge spillover theory shows how entrepreneurship can contribute to industry growth by helping knowledge to spill over or to permeate the filter that impedes knowledge spillover.

The following two pooled OLS regressions are used for empirical analysis

$$I_{it} = \alpha + \beta_1(K_{it}) + \beta(E_{it}) + \beta_3(K_{it}E_{it}) + \beta_4(\text{controls}_{i,t}) + \beta_5(\text{Years,t}) + e_{it}$$

where I is either "new-to-the-market innovation" (the share of turnover attributable to new or significantly improved products that are new to the market) or "new-to-the-firm innovation" (the share of turnover attributable to new or significantly improved products that are new to the firm); K denotes the rate of knowledge-intensive firms measured by the share of firms that applied for at least one patent in the last three years; E denotes the business ownership rate as a proxy for entrepreneurship rate; "Controls" denote the control variables which are the natural logarithm of GDP and GDP per capita; "Years" correspond to year dummies for the years 1998, 2000, 2004, and 2006; and t and i are year and country indices, respectively.

The empirical data are from the Community Innovation Survey (CIS), the COMPENDIA database, and the OECD Economic Outlook Database. Twenty-one European countries (Austria, Denmark, Netherlands, UK, Sweden, etc.) and 57 observations are used over the period 1996–2006. The descriptive statistics of the data are as follows:

Variables	Mean	Median	Standard deviation
1. New-to-the-market innovation (%)	8.12	7.30	3.96
2. New-to-the-firm innovation (%)	12.84	10.40	7.71
3. ln GDP	12.47	12.25	1.17
4. GDP (mill. $)	486,722	208,854	565,727
5. ln GDP per capita	10.03	10.08	0.31
6. Entrepreneurship rate (%)	10.78	9.80	3.90
7. Rate of knowledge-intensive firms	10.14	9.70	6.38

Table 6.2 shows the results of pooled OLS regressions with respect to new-to-the-market innovations. The empirical analysis is conducted in four steps with four representative models. Model I is the baseline model which includes the macroeconomic control variables and the year dummies. This model explains 13 %

Table 6.2 Pooled OLS regression coefficients on new-to-the-market innovations

	Model I	Model II	Model III	Model IV
1. ln GDP	0.28	−0.21	−0.34	−0.70
2. ln GDP per capita	−3.17**	−6.54**	−6.69**	−7.38**
3. Rate of knowledge-intensive firms	–	–	0.09	−0.38**
4. Year dummies				
2000	1.94	2.94	3.01	3.25**
2004	1.70	2.81**	2.91**	3.13*
2006	2.64**	3.98**	4.16**	3.13*
F-value	5.31**	4.67**	4.60**	6.53**
R^2	0.13	0.22	0.22	0.29
N	57	57	57	57

Note: One and two asterisks denote 5 % and 1 % significance of t-values respectively

Table 6.3 Pooled OLS regression coefficients on new-to-the-firm innovations

	Model I	Model II	Model III	Model IV
1. ln GDP	1.91	0.98	1.40	1.52
2. ln GDP per capita	2.50	−3.89	−3.41	−3.19
3. Rate of knowledge-intensive firms	–	0.51*	0.39**	0.59
4. Year dummies				
2000	−0.95	0.96	0.73	0.65
2004	−9.53**	−7.44**	−7.78**	−7.85**
2006	−10.43**	−7.88**	−8.48**	−8.53**
F-value	6.51*	6.93**	6.13**	6.84**
R^2	0.50	0.59	0.60	0.61
N	57	57	57	57

Note: One and two asterisks denote 5 % and 1 % significance of t-values respectively

of the variation of the dependent variable: new-to-the-market innovation. In Model II, we see that a higher share of knowledge-intensive firms leads to a higher share of new-to-the-market innovations (i.e., $\beta_1 = 0.27$). Model III includes the entrepreneurship variable in the model and shows that the rate of entrepreneurship itself does not seem to affect new-to-the-market innovation. Model IV tests the moderation effect of entrepreneurship, and the interaction term (omitted in the table) shows a positive effect (i.e., $\beta_3 = 0.07, p < 0.05$). A higher rate of entrepreneurship seems to increase the rate by which knowledge leads to new-to-the-market innovations, indicating that a higher rate of entrepreneurship facilitates the commercialization of knowledge.

Table 6.3 shows overall that knowledge creation clearly leads to a higher share of new-to-the-firm innovations. It also shows that entrepreneurship does not have an effect with regard to new-to-the-firm products. These results may imply that public policy should be directed more to the new-to-the-market innovations so far as the spillover benefits are concerned.

The tables also indicate a positive time trend for new-to-the-market innovations and a negative time trend for new-to-the-firm innovations. The ratio of "time" innovation vs. imitative innovation has increased over time in the 21 European countries. This phenomenon is one of the main indicators of the switch from a "managed" to an "entrepreneurial" economy. To check the robustness of the estimated results the authors estimated both random and fixed-effects models (not reported here). These estimates confirm the main finding that entrepreneurship moderates the relationship between knowledge and new-to-the-market innovations.

We conclude this section by noting that from an innovation policy perspective promoting the production of new knowledge by means of R&D subsidies or university education is not sufficient. It is equally important for entrepreneurs to turn this new knowledge into innovative products to fuel economic growth. A more appropriate long-term strategy for the policymakers would be to promote entrepreneurship education to increase the number of qualifies and risk-taking entrepreneurs.

6.5 Innovation Policy Changes

Modern innovations occur in many forms. Besides Schumpeter's analysis of six types of innovations, two of the important ones are rivalrous innovation and coinvention through endogenous innovation. In nonrivalrous innovation the firms cooperate to take advantage of economies of scale and scope. The spillover effects of different firms' R&D are jointly utilized and the overall impact would be welfare increasing. In rivalrous competition, however, the race for winning the innovation for new processes or new products continues. Successful innovations arise as a result of a Poisson process with an intensity u. The probability that a firm's innovations are successful during the period $\mathrm{d}t$ is $u\mathrm{d}t$. Since firms are assumed to have equal chances, the probability that any particular firm becomes a winner in the race is $(u/n)\mathrm{d}t$, where n is the number of firms. The expected monopoly surplus from winning the R&D innovation race is then $(su/n)\mathrm{d}t$, where s is the monopoly surplus due to quality improvement due to innovation and the resulting price rise. Denoting variable cost by $v(u)$ and the R&D fixed cost by f, the firm's instantaneous expected profit is then

$$\pi(u,n) = (su/n) - v(u) - f$$

If we assume $v(u) = u^a/a$, then the optimal R&D intensity $u(n)$ maximizing the profit function above becomes

$$u(n) = (s/n)^{1/(a-1)}$$

This yields the optimal profit function

$$\pi(n) = a^{-1}(s/n)^a - f$$

This shows that the optimal profit declines monotonically as n increases. In rivalrous innovation the firms are likely to be either Cournot–Nash competitors or Stackelberg rivals (i.e., leader–follower). In the former case a firm's payoff from innovation depends on the number of other firms that innovate successfully. As the number of successful firms rises each firm's sales fall and so does the benefit of successful R&D investment. This effect leads each firm to do less R&D as the number of firms rises. Also, each firm can benefit from successful R&D, even if other firms succeed as well. This effect can raise the total industry incentives to conduct multiple R&D projects. The state subsidy can help in fostering cooperation in R&D project through information network. This might help in raising total industry R&D. Many successful Southeast Asian countries which achieved high rates of industry growth followed such innovation policies.

The Spence model we considered before emphasized the challenges of the appropriability problem of an innovating firm, which is large and concentrated and incurs large R&D investments as fixed costs. This investment reduces unit costs and improves productivity, but the spillover benefits cannot all be internally appropriated by the successful innovative firm. This tends to reduce the initial incentives for innovation. But the output price of the successful firm closely equals to marginal cost, which may closely equal a negligible amount or even zero. This marginal cost however is the cost of transmitting to other firms. Thus there arises an unpleasant trade-off between incentives and the efficiency due to unit cost reduction. The most direct way to deal with this problem as suggested by Spence (1984) is to subsidize the R&D activity of the innovator or provide a mechanism for R&D cooperation. This type of remedy for innovation challenge tends to make the industry level more efficient and optimal from a social welfare standpoint.

Another policy problem arises when the R&D investment for process innovation occurs in a Cournot duopoly framework, where firms may either undertake independent ventures or form a cartel for R&D investments. In the short run unit costs decline due to R&D and its spillover, but in the long run unit costs may rise due to fixed R&D capital subject to diminishing returns. Cellini and Lambertini (2009) have formulated this type of Cournot model and compared the social welfare effects of these two settings, i.e., independent R&D laboratories and joint labs as cartels. They showed that cartelization dominates competition between independent ventures. The cartel framework is found to be more beneficial from a social welfare viewpoint and hence it justifies appropriate state policies to foster such cooperation for forming an R&D cartel.

These models however tend to underemphasize one factor that is the voluntary dissemination and exchange of knowledge of technology and information structures. If a firm supplies its technology and R&D information to a rival and the favor is reciprocated, then both firms end up more strengthened in respect of a third competitor. Thus market forces in this framework tend to provide a strong incentive for the formation of informed technology and information consortia. Baumol (2002) emphasized this aspect in analyzing the growth miracle of capitalism in recent decades through the worldwide diffusion of the innovation upsurge in software, telecommunications, and bioengineering fields. In many successful NICs of Southeast Asia like China, South Korea, and Taiwan, industrial parks, export zones, and technology consortia have been deliberately sponsored by the state as a sharing center of new knowledge about the latest technology and software. This may provide one the most economic ways of solving the externality and spillover challenge.

6.6 Innovation Experience in Asia

Why is innovation so important for industry growth under private capitalism and what can the state and private industry do to improve it? Innovation challenges have been successfully handled by many Southeast Asian countries like China, Taiwan, South Korea, and Singapore which achieved a high rate of economic growth in the last three decades. Many economists have called this as Asian growth miracles. Two sources of this miracle are identified by the World Bank Report (1993): accumulation of capital, both physical and human, and the significant productivity gains. The contributions of innovations through human capital are very important here. Productivity change (or gains) is measured by total factor productivity (TFP) growth, which is estimated in a simple neoclassical framework by subtracting from total national output growth, the portion of growth due to both physical and human capital accumulation, and labor force. The following estimates for TFP over the period 1960–1989 are from the World Bank Report:

Hong Kong	3.647	Indonesia	1.254
Singapore	1.190	South Korea	3.102
Japan	3.478	China	3.560
Latin America	0.127	Africa	−0.998

The innovation and R&D investment policy experiences of three countries, China, Taiwan, and South Korea, deserve some comments since they show how the innovation challenges were handled by the rapidly growing economies which performed the so-called growth miracles.

Recently Wu (2008) decomposed sources of economic growth by TFP, technical efficiency (TE), and scale efficiency as follows:

Several points emerge here. Overall the TFP growth accounts for about 27 % of the total growth of China over the period 1993–2004. Wu's conclusion is that there is further scope for gains in TFP in the recent period 2005–2010. Later studies have confirmed this. Secondly, the contribution of ICT capital turns out to be 4.37 during 1992–2000 and 2.65 during 2001–2004. The role of incremental innovation can be

Table 6.4 Sources of growth in China (annual average rate)

	1993–1997	1998–2000	2001–2004
TFP	1.64	4.30	3.56
TE	−0.26	1.89	1.19
SE	1.16	0.62	0.80
Output	12.40	8.99	10.95
TFP/output	13.23	47.81	32.47

Table 6.5 National innovative capacity index (2002–2003)

Country	2002	2003	Change
USA	37.21	26.60	−0.61
Singapore	32.45	34.19	1.74
Taiwan	32.34	32.84	0.50
South Korea	30.59	31.13	0.54
China	26.01	25.86	0.20
India	25.24	25.52	0.28
Japan	33.98	34.62	0.64

indirectly detected through ICT capital investment. A study by Qian and Smyth (2006) treats human capital as a separate input and shows that TFP and human capital contribute 22 % and 13 % of total GDP growth respectively.

One should mention here about China's recent policy announcement to push China up the innovation ladder in the next 15 years 2010–2025 by raising R&D spending from 1.3 % in 2009 to 2.5 % in 2020. This will put China on par with the USA and Germany. Secondly, the Long-Term Science and Technology Development Plan for the next 15 years has stipulated several long-term efforts. One is to raise productivity from technology improvement to 60 % by 2020 from 39 % in 2009. Another is to set up and improve the record of technology-based patents, where Taiwan ranks close to the top in the world along with the USA.

To measure innovation intensity Porter and Stern (2004) developed an index termed ICI (innovative capacity index), which comprises technology innovations and R&D. Table 6.5 provides a brief summary of findings for the period 2002–2003.

Clearly Taiwan's rank is higher than China and South Korea and is more comparable to Japan. What are the major sources of this significant innovative potential for Taiwan? First of all, electronics has been a major driving force and the innovation has played a catalytic role. In recent years 2004–2010 Taiwan has become more R&D intensive. New areas are developing where Taiwan appears to be rapidly approaching the cutting edge of technology such as wireless integrated circuits and design of new inventions. Secondly, if we measure innovative capabilities by the amount of industrial patenting and use the number of patents granted in the USA as the metric, Taiwan looks very strong. In per capita terms Taiwan and Israel are the only two emerging economies to close the gap with the G7 countries in terms of the patent per capita ratio, with Taiwan being next after the USA and Japan. Note that the public policy fostering diffusion of R&D knowledge has played a very catalytic role here. Public laboratories like the ITRI (Industrial Technology Research Institute) continue to be the major sources of patents besides branding, design, and

Table 6.6 US patent awards (1995) to foreign countries: top 11 countries

Country	Invention	Design
Japan	21,925	1,149
Germany	7,311	258
France	3,029	234
UK	2,425	194
Canada	1,964	240
Italy	1,271	172
Switzerland	1,196	93
Taiwan	1,000	250
Netherlands	855	66
Sweden	627	98
South Korea	538	48

incremental innovations. Also many innovation-oriented companies often enter into partnership with China and foster sharing of technical knowledge and innovation designs. Third, Taiwan has recently made intensified efforts to develop critical masses of leading researchers and engineers. It has strengthened closer direct links between public universities and industry by making Taiwan's electronics, computers, and communication sectors along with manufacturing and services. As we mentioned before Taiwan's record of performance in the IT (information technology) sector is most impressive. In terms of the average number of annual US patents per million people, the top ranking in the world in 2004 are : 1 for the USA, 2 for Japan, and 3 for Taiwan.

Two elements of the innovation dynamics are most important for the rapid growth attained by the NICs in Southeast Asia. One is the openness in international trade involving foreign direct investment. This allows the spillover effects of innovative investment in the USA to be captured by the successful NICs in Asia like Taiwan, South Korea, Hong Kong, and Japan. The second is the learning-by-doing effect of human capital element in innovation in the form of education, skills, and research knowledge. Two of the successful NICs in Asia need to be specially mentioned here. One is South Korea and the other China including Hong Kong. These two countries differ in several ways. For example, Korea is democratic, whereas China is not. Korea follows market capitalism with much less control and regulation by the state. Also Korea is more open to global trade unlike China and Korea's exchange rate is determined in open markets unlike China.

Korea's growth episode resembles the pattern of other successful NICs in Asia. Over the period 1985–1995 the annual growth rate of income per capita has exceeded 7.6 % accompanied by high investment rate exceeding 7.1 %. Productivity and efficiency gains provided the key sources of growth. The private sector played a powerful role in both domestic and foreign trade. South Korea's export boom amounted to about 35 % of GDP during the 1990s and most of it was in nontraditional goods such as color TV, electronics, and cars. The share of IT products and services which are innovation intensive in total merchandise exports through 1980 to 1989 has grown from 10 in 1980 to 22 in 1989 and rose higher than

Table 6.7 Average annual growth rate (%) in Korea

	1981–1985	1996–2000
GDP	7.5	4.7
IT Capital	0.2	0.4
Labor	1.07	0.37
TFP	4.36	0.39
IT contribution	0.66	8.4

30 in 2010. The IT sector in Korea has helped greatly in the process of technology and innovation diffusion. The role of IT investment in Korean economic growth is as follows (Table 6.7):

These estimates from Kim (2002) provide indirect evidence of the impact of innovation through IT capital and TFP. Kim's analysis showed that the IT sector investment has helped diffuse productivity in other sectors of the economy. Like Taiwan, R&D knowledge in public universities and public laboratories have been systematically transferred to the private business sector. The state subsidy program has directly helped in this knowledge diffusion process.

The trend of growth miracles in Southeast Asia strongly emphasizes the role of successful innovations in the IT sectors. Learning by doing, knowledge diffusion, and worldwide competition all have carried the banner of this innovation. An example of another successful incremental innovation would be helpful here. In the late 1960s Kelleher and King went to found Southwest Airlines, inventing low-cost air travel by using secondary airports and flying passengers directly to their destination. They cut out free meals, passing on the savings to their customers. Little did they know at the time that Southwest Airlines was making innovation history. A highly innovative company entered the air travel market and changed the very market itself. This is exactly what Schumpeter had in mind when he developed the broad concept of innovation. Growth miracles of Southeast Asia bear this testimony.

Chapter 7
Managing Innovation

Management is the essence of success in innovation. Viewed as an R&D process innovation involves optimal decision making under risk and uncertainty. Large-scale economies associated with modern innovations have significant impact on widening the market thus inviting intense competition. Rivalry in competition and significant spillover effects of information and research knowledge create a state of hypercompetition as discussed by D'Aveni and others. Schumpeter's notion of creative destruction implies creativity attributed to the innovative process, which destruction is at its core. In Schumpeterian framework the innovation coming from within has a destructive side. Along with pioneering entrepreneurs there are laggards, "mere managers" in Schumpeter's description who cannot keep up with the pace of change brought about by new combinations of products, new processes, and innovative organizations. Product innovation may lead to the emergence of new industries thus causing the demise of existing ones.

Two other developments are important in the managerial perspective. One is the need for a flexible organization structure sometimes called the corporate lattice model, when in this new information, age process and product changes are so frequent. Secondly, the economies today have undergone a transformation from large-scale material manufacturing to the design and use of new technologies and new communication processes, which are increasingly characterized by increasing returns (IR). These are mechanisms of positive feedback which reinforce the success once achieved. They occur due to four reasons: (1) high fixed costs with very low marginal costs; (2) large network effects where the product/service value increases with the number of users; (3) high switching cost, which helps to lock in consumers; and (4) information-intensive aspects which have large learning-by-doing effects.

We discuss here three aspects of the managerial problems associated with innovation processes:

1. Management challenges
2. Business under increasing returns
3. Digital divide and innovations

J. Sengupta, *Theory of Innovation: A New Paradigm of Growth*,
DOI 10.1007/978-3-319-02183-6_7,

These aspects are most important for industry growth and decline in today's information-intensive world.

7.1 Management Strategies

Schumpeter introduced two sides of innovation: one is creative and the other destructive. The first brings in new entry of firms causing industry growth. The second initiates exit from the demise of obsolete industries. He summed up the process of "creative destruction" as follows:

> The opening up of new markets, foreign or domestic, and the organizational development from the craft shop and factory to such concerns as US steel illustrate the same process of industrial mutation, that necessarily revolutionizes the economic structure from within, necessarily destroying the old one, necessarily creating a new one. The process of Creative Destruction is the essential fact about capitalism. It is what capitalism consists in and what every capitalist concern has got to live in. (Schumpeter 1942, p82–83)

In Schumpeter's theory of growth the economic impact of technical change is considered a disequilibrium phenomenon. Thus in a capitalist economy characterized by ongoing diffusion processes of innovations, time averages are more important than steady-state values even in a long-run perspective. Having emphasized the Walrasian result that entrepreneurial profits are absent in a stationary process, Schumpeter discussed how successful innovations usually translate into profits in an evolutionary process: innovations often consist in changes in the productive process in the widest sense, the aim of which is to produce a unit of product with less expense and thus create a discrepancy between their existing price and their new costs. Schumpeter anticipates here the more recent distinction between "lower costs" and "differentiation" as emphasized by Porter (1990) and others as the two basic types of competitive advantage. These usually translate into higher productivity and Schumpeterian innovation attempts to ensure this through disequilibrating mechanisms. Of course many attempted innovations fail to this standard, but then they are not likely to succeed in market competition.

We now discuss some of the important challenges for management in the innovation framework as follows:

1. Economic evolution and the process of competition
2. Developing employee engagement that drives high performance
3. Aligning projects with company objectives in unpredictable times

Schumpeter discussed in detail how the economic system responds to the intrusion of an innovation. For some of the old firms new opportunities for expansion open up; this indicates that innovations have important positive externalities and spillover impact. But for others the emergence of new methods means economic demise. They are forced to undergo a painful process of modernization, rationalization, and reconstructions. The managers have distinct roles in these two environments. For innovative ventures to succeed they have to align the R&D

projects to the firms' objectives and this requires creativity and optimal management of R&D funds. For noninnovative firms cutting losses sooner and starting a new process of reconstruction are the best strategies. Darwin's theory of genetic evolution explains the process of endogenous change by the interaction of several fundamental mechanisms. The basic tool is replicator dynamics which applies the selection theory of evolution where the frequency of a species (which may represent new innovation processes in Schumpeterian theory of economic evolution) grows differentially according to whether it is below or above the average fitness. If the concept of fitness in genetic theory is replaced by economic efficiency or core competence in organization theory, the replicator dynamics in firm growth can explain the industry evolution. One simple formulation of replicator dynamics is of the differential equation form

$$\dot{x}_i = \mathrm{d}x_i/\mathrm{d}t = Ax_i(E_i - \bar{E}), i = 1, 2, \ldots, n$$

where $\bar{E} = \sum x_i E_i$ is average firm fitness, E_i is fitness of species (or firm) i in a population of n interacting species, and x_i is the proportion of species i in the population. In economic evolution x_i may represent for example the proportion of firms using technology or innovation and E_i may denote economic efficiency. The survival of the fittest principle implies that the firms with the largest level of fitness E_m exceeding the average $\bar{E}$ will succeed in the innovation competition. Kelm (1997) has discussed in some detail the Darwinian mechanisms in the Schumpeterian model of economic evolution. In Darwinian theory the interaction of three mechanisms operating through individual members (or firms in economic evolution theory) of the population:

1. A mechanism of *information storage* by which some relatively stable characteristics (or core competence) are preserved over time
2. A mechanism of *endogenous change* which generates new variations continually
3. A mechanism of *selective retention* by which the frequency of some variations relative to others is increased

Schumpeter recognized not only the fundamental importance of innovation but also its difficulty, which stand out if innovation is contrasted with ordinary business routine. The ultimate source of this difficulty is bounded rationality. The inevitability of error is due to the impossibility of surveying exhaustively all of the effects and countereffects of all innovation projects. Schumpeter anticipated here the concept of "genuine uncertainty." That is why he emphasized the phenomenon of *entrepreneurship* as the special explanation of the successful innovation mechanism. His discussion of entrepreneurship as the pivotal source of endogenous change serves to highlight the fundamental difference between economic and biological evolution. In biology the mutation as the source of endogenous change is random (i.e., not adaptively directed). Despite the constraints of genuine uncertainty, entrepreneurs do not change routines randomly; they put expected profit at a positive level as the premium obtainable upon successful innovations. Since innovation is the fundamental source entrepreneurial profit and profit is required for survival, failure to innovate is the fundamental reason for competitive elimination. Entrepreneurship is the fundamental source of endogenous economic change and learning is the mechanism of selective retention by which the innovations created by entrepreneurs are spread, but the incentives that drive these conscious human actions are generated by the competitive process involving differential survival of firms. The characteristic effects of different types of innovation on the structure of industry growth and decline reflect the operation of the different Darwinian forces which are implicit in Schumpeter's theory. For example, incremental innovations are closer to existing routines and offer more scope for conscious *ex ante* evaluation than radical innovations do.

Computerization, the Internet, and other ICT innovations have increased the sophistication of modern production technology with complex and diverse implications. Changes in modern technology such as the development of computer-aided design and manufacturing (CAD/CAM) have changed traditional ideas of price/quality trade-offs and allowed the production of high-quality, tailor-made goods and services at low cost. In using these new technologies however the managers in the 2000s must choose and choose rationally between reformulating their strategies and using these technologies incrementally to reinforce the traditional modes of production and organization.

Benko and Anderson (2010) have recently emphasized the role of "the engagement principle" in building a corporate lattice model of organization structure for achieving high performance in the changing world of work and management. Engagement principle describes the extent to which employees are motivated to perform at a high level and to advocate for their company's products and services externally to other contacts in their network. Numerous studies have documented the benefits of engagement. Companies with high engagement scores boast the following:

(a) Improved shareholder value: Firm's with top-quartile employee engagement realized 2.6 times the earnings per share growth as firms with below average engagement.
(b) Higher returns on assets: Companies in the top quartile of employee engagement experiences double the return on assets of those in the lowest quartile.
(c) Higher quality: Health-care facilities with engagement scores in the top 50 % had medication error rates as much as 25 % lower than others in the bottom half.
(d) Higher profitability and productivity: Asian companies with high engagement scores were 40 %.

The corporate lattice model provides a broad perspective on talent development among modern day managers who have the constant need to develop creativity and adapt to winds of constant change. This model moves companies from a ladder management view centered strictly on only vertical progression. It formalizes opportunities for managerial workers to progress horizontally. This helps develop the breadth and depth of capabilities that they need to compete. A major drawback of the ladder model of corporate management is that it does not adequately develop specialists with creative knowledge who may have deep expertise, institutional knowledge, nor does it readily adapt when new types of skills emerge. The lattice model provides career pathways for managers and skilled workers, the pathways which provide flexibility, enabling knowledge specialists to customize their journey over time and companies to deepen their pool of expertise.

The lattice model implies a lattice leadership development mechanism. Benko and Anderson provide a successful example. As collaboration through councils and boards became the management norm, Cisco realized the need to develop leadership talent differently. It had to evolve the traditional trajectories of executive careers to align with its new model. Cisco is now moving high potential executives laterally as well as vertically to build the breadth of business perspectives that collaboration requires.

Aligning projects with company objectives is of the critical managerial change in the technology-intensive world of business today. Between 1997 and 2001, about $2.5 trillion was spent on technology in the USA, nearly double the amount for the previous five years. At the same time Morgan Stanley reports that the IT capital spending in the USA has grown approximately 50 % of nominal US business capital spending, and this trend is continuing.

The alignment principle for the managerial worker emphasized flexible adaptivity, which has three components: aligning the companies project portfolio with its objectives, aligning the projects in the portfolio to each other, and aligning the portfolio and the company's objective with the changing realities of current business today. Recently Benko and Carolean (2004) have characterized today's changing and volatile business world as the new information frontier. They argue that we are just beginning to take advantage of the exciting new opportunities afforded by the information frontier, e.g., the Internet, social network, computer revolution, and miniaturization of communication technology. Four characteristics of this information frontier have been identified by Benko and Carolean:

1. Organizational transparency
2. Velocity of information speed
3. Reduced transactional friction
4. Role blurring

Easy availability of information and data has become much easier now to look into organizations. In the past upper-hand managers could shape and send information on a "need to know" basis. Today it is more likely that everyone in and around the organization can get plenty of data through Internet, social network, and other channels. The manager should be aware of this transparency, which puts a harsh spotlight on a company's weakest links. Velocity refers to the speed at which information flows today. Managers have to be increasingly aware of this situation when they have to take more decisions under this threat. A third feature of the new information frontier is the opportunity for reduced transactional friction. Since data network has been more robust, more efficient, and more standardized today, the costs of contracting, communicating, and coordinating have been considerably reduced. The transaction cost theory of Coase, North, and others implies that this aspect of information frontier would considerably facilitate market demands and expand markets. This reduction in transaction cost has blurred the role of different agents to the transaction. In today's economy for example the supplier may be a part owner, part competitor.

Living on the information frontier requires companies to act, respond, and invest precisely at the greatest moment of uncertainty. Managers have to develop skills to respond under the condition of risk and uncertainty. Need for good decision makers is all the more great today.

7.2 Business Under Increasing Returns

Recent innovations in communications and information technology have altered the practice of business organization today. These have created mechanisms of positive feedback so that there is a dramatic shift to economic activity that is increasingly characterized by increasing returns (IR). Knowledge capital and R&D investment with large fixed costs but with low variable costs have generated this framework resulting in lower prices and larger markets. Two characteristics are most important here. One is the more rapid obsolescence of knowledge in IR industries than in traditional DR (diminishing returns) industries. The typically short life cycles of products/services and processes make a constantly renewed stock of proprietary knowledge a critically important asset. Secondly, in IR activities, recent upsurges in information availability and network interconnections are breaking down the isolating mechanisms that prevented rivals before from access to proprietary knowledge.

Recently Nachum (2002) estimated a regression model for assessing the impact of several firm-specific advantages such as innovative capabilities, scale effect, and flexibility of organizational structure on the dependent variable: the propensity of firms competing in IR or DR industries (this is measured by outward FDI). The model is of the form

$$y(it) = \beta_0 + \sum_{j=1}^{m} \beta_j A_j(it) + e(it)$$

where $y(it)$ = FDI (i.e., total capital flow of industry i in year t)

A_j = firm specific advantages

e = random error term

N = no. of observations: 390 for IR and 260 for DR industries

The model is estimated for the years 1989–1998 based on outward FDI (foreign direct investment) panel data from the USA. Data include 390 IR and 260 DR industries. The IR industries include the advanced knowledge-based industries such as electronic components and accessories, industrial chemicals, and pharmaceuticals.

The DR industries include the traditional industries such as primary metal, meat products, and grain mills. The firm-specific advantages are as follows:

A_1 = innovative capabilities measured by R&D expenditure as share of sales

A_2 = scale measured by total sales

A_3 = differentiation measured by advertising expenditure as share of sales (promoting brand royalty)

A_4 = cost of sales

A_5 = multinationality advantages measured by affiliates'

A_6 = risk measured by credit and liquidity risk

A_7 = entrepreneurship measured by growth of international sales

A_8 = flexibility of organizational structure measured by sales of affiliates to parent companies

A_9 = networking measured by the size of affiliates in terms of volume of employment: firm size

The estimates are as follows:

	(IR)	(DR)	Difference Statistic
β_0	110.40	103.10	–
β_1	15550.9***	59753.4*	−9.8***
β_2	0.199**	0.313**	−0.210**
β_3	−1987.3	2325.4	−4861***
β_4	23646.4	122375.8*	10375.4
β_5	49710.5***	83031.4	22724.4
β_6	9432.5*	−4643.4	−8.68***
β_7	28346.0**	13457.8*	7639.6
β_8	−26247.3*	−117981.4	6459.3
β_9	−1.073	3.88	3.78
N	390	260	
Adj R^2	0.755	0.497	

Here one, two, and three asterisks denote significance of t values at 5 %, 1 %, and 0.1 %, respectively. Difference statistics are calculated to measure whether the explanatory variable has a differential impact on international expansion in IR and DR industries.

Several economic implications of the statistical results above are to be noted. First of all, the strong explanatory power of the innovation capabilities confirms the vital role of innovation in these knowledge-intensive and rapidly changing technologies. Secondly, the significant explanatory power of entrepreneurship, flexibility, and networking highlights the importance of these attributes in IR industries. Finally, a comparison of the IR and DR models supports the following hypothesis strongly:

Flexibility of organizational structure has a significant positive impact on IR industries with a large network of global trade and this impact is stronger than in DR industries.

Many of the important lessons for managers in this modern age may be assessed. Dynamic capability theory of management emphasizes the strategy of adaptivity of human capital. Adaptive human capital not only facilitates the adoption of more advanced technology but also makes it easier to innovate at the technology frontier. Another facet of organizational flexibility that the modern manager should be aware of is the lattice structure of R&D investment strategy. For this purpose the innovative firms to attain success should be organized as multidimensional or M-form where the CEP is at the top with a functional decentralization into manufacturing, sales, distribution finance, and R&D divisions. According to Williamson (1985) it is the most significant organizational innovation of the twentieth century. The essential feature of the M-form is that the profit responsibility is decentralized to the level of individual product lines, individual brands, and individual geographic markets. An example of success of this organizational form is Pilkington Glass company, which spend 8 years and 21 million dollars to prefect its float plate glass process. This innovation sustained Pilkington's dominance in the glass industry.

Managers in today's business world have to be aware that learning by doing and increasing returns to knowledge and information capital are vital to modern business as an evolutionary process, where firms initially gain competitive advantage by altering the basis of static competition. The successful innovators win not just by recognizing new markets or new technologies but also by moving aggressively to exploit them. They sustain their first mover advantages by investing to improve the existing sources of advantages and to create new ones.

7.3 Digital Divide and Innovations

Digital divide refers to the technology gap that remains today in the business world. In many third world countries the technology gap persists through direct or indirect political pressures. The technology gap model predicts that the countries leading in technology and innovations excels in growth of productivity and of demand and the follower economy lags behind. The rapid growth of NICs in Southeast Asia over the last 3 decades has confirmed this trend. Indirectly this supports the Schumpeterian hypothesis of creative accumulation, which complements his concept of creative destruction.

As competition intensifies in today's information-intensive frontiers the challenge of innovation efficiency would be all the greater. The digital divide would be all the more present. Recently Lopez-Claros and Mata (2010) have constructed a composite measure called the innovation capacity index (ICI) based on a weighted average of four pillars as they are called:

1. Institutional environment which is efficiency in public sector management, corruption perception index, and the state of the macroeconomy
2. Human capital comprising adult literacy, primary, secondary, and tertiary enrollments
3. R&D which includes R&D worker density and patents and trademarks
4. Use of information technologies

This ICI index in its 2009 version covers 131 countries. Some ranking of this index over 2009–2010 is as follows:

Country	ICI rank	ICI score
Sweden	1	82.2
USA	2	77.8
Singapore	6	76.5
Taiwan	13	72.9
Japan	15	72.1
Hong Kong	16	71.3
South Korea	19	70.0
Malaysia	34	57.3
China	65	49.5
India	86	45.6
Brazil	87	45.2

Clearly the NICs in Asia fare very well in the ICI rankings. The incentive to innovate has played a most dynamic role in stimulating industry growth in the fast developing NICs in Asia.

Growth of modern high-tech industries today is increasingly playing a cumulative accelerator-type role in the overall growth of an economy. The world market today belongs to those firms who compete and succeed in innovations. Managers today have to take a leading role to follow and accelerate the information frontier. Technology today is largely driven by the information frontier and innovation is at the center of this information technology frontier. Innovation introduces new efficiency into the industrial structure by following two dynamic strategies. One is to introduce rivalrous competition, when the first mover enters the market with new innovations and sustains the competition successfully. The second is to pursue dynamic efficiency over time and continually sustain it through incremental innovations and R&D investment. Four types of dynamic efficiency are the most important for the managers to pursue:

1. Production efficiency: This involves making continual investments to lower future costs, improving future quality, and creating know-how for the long run.

2. Innovation efficiency: Racing up the escalation ladder in timing and knowledge frontier.
3. Resource efficiency: Here firms look for assets in the whole industry that are currently underutilized. A good example is Oracle, which has recently bough several companies like Peoplesoft and Sun's data division to complement its existing resource base and expand successfully.
4. Access efficiency: Racing up the success ladder in the stronghold arena leads to this efficiency. Exploring large economies of scale in the globalization of markets provides the critical source of this efficiency.

In this dynamic market the concept of Walrasian equilibrium lose much of its usefulness. Companies today pursue vigorous rivalrous competition to keep new entrants out of the industry by aggressively moving forward in the race for foraging and sustaining the four types of dynamic efficiency as above. These efficiency concepts go far beyond the traditional notions of competitive efficiency. They also go beyond the Schumpeterian view of innovation efficiency alone. Today's goal for industry growth is ingenuity and managerial skills. Success belongs to the smartest, the most knowledge seeking, and the most innovative CEOs!

Bibliography

Abernathy, William J and James M Utterback, "Patterns of industrial innovation," *Technological Review*, 1978, pp. 41–47.

Acs, Zoltan J., Pontus Brodde Braunerhjelm, David Audretsch, and Bo Carlsson, "The knowledge spillover theory of entrepreneurship," *Small Business Economics*, 2009, *32* (1), 15–30.

Agliardi, Edward, *Positive Feedback Economics*, New York: St. Martin's Press, 1998.

Andersen, E.S., *Joseph A. Schumpeter: A Theory of Social and Economic Evolution*, New York: Palgrave Macmillan, 2011.

Antonelli, Charles, "The diffusion of new information in economic growth," *Journal of Evolutionary Economics*, 2003, *10*, 34–46

Arrow, Kenneth, "Economic welfare and the allocation of resources for invention," in "The rate and direction of inventive activity: Economic and social factors," Princeton: Princeton University Press, 1962, pp. 609–626.

Azoulay, Pierre and Rebecca Henderson, "Relational contracts and tapered integration: Evidence from drug development," *Unpublished working paper*, 2001.

Barras, Richard, "Interactive innovation in financial and business services: the vanguard of the service revolution," *Research Policy*, 1990, *19* (3), 215–237.

Barro, Robert, "Government Spending in a Simple Model of Endogenous Growth," *Journal of Political Economy*, 1990, pp. S103–S125.

Baumol, William J, *The Free Market Innovation Machine: Analyzing the Growth Miracle of Capitalism*, Princeton: Princeton University Press, 2002.

Benko, Julian and S Carolean, "Schumpeterian ideas," *Applied Economics*, 2004, *115*, 25–49.

Benko, Cathleen and Molly Anderson, *The Corporate Lattice: Achieving High Performance in the Changing World of Work*, Boston, Mass: Harvard Business Review Press, 2010.

Bernstein, John and Nadiri, "Michael Inter industry R & D spillovers, rates of return and production in high-tech industries," *American Economic Review*, 1988, *78*, 429–434.

Besanko, David, David Dranove, Mark Shanley, and Scott Schaefer, *Economics of strategy*, 5th ed., New York: John Wiley, 2010.

Block, Joern H, Roy Thurik, and Haibo Zhou, "What turns knowledge into innovative products? The role of entrepreneurship and knowledge spillovers," *Journal of Evolutionary Economics*, 2012, *22*, 191–210.

Boyd, Robert and Paul Richardson, *Culture and Evolutionary process*, Chicago: University of Chicago Press, 1985.

Braunerhjelm, Peter, Z Acs and B Carlsson, "The missing link: knowledge diffusion and entrepreneurship in growth," *Small Business Economics*, 2010, *34*, 104–125.

J. Sengupta, *Theory of Innovation: A New Paradigm of Growth*,
DOI 10.1007/978-3-319-02183-6,

Caballero, Ricardo J and Adam B Jaffe, "How high are the giants' shoulders: An empirical assessment of knowledge spillovers and creative destruction in a model of economic growth," in "NBER Macroeconomics Annual 1993, Volume 8," Massachusetts: MIT press, 1993, pp. 15–74.

Cellini, Roberto and Luca Lambertini, "Dynamic R&D with spillovers: Competition vs cooperation," *Journal of Economic Dynamics and Control*, 2009, *33* (3), 568–582.

Christensen, Clayton, *The innovator's dilemma: when new technologies cause great firms to fail*, Boston, MA: Harvard Business School Press, 1997.

Cohen, Wesley M and Daniel A Levinthal, "Innovation and learning: the two faces of R & D," *The economic journal*, 1989, *99* (397), 569–596.

Cowan, John and William Ozman, "Technical creativity and economic change," *Metroeconomica*, 2004, *105*, 25–32.

D'Aveni, Richard, "Hypercompetition: Managing the dynamics of strategic management," *New York: Free Press*, 1994.

d'Aspremont C and A Jacquemin, "Co operative and non co operative R & D in duopoly with spillovers," *American Economic Review*, 1988, *78*, 1133–1137.

DeBresson, Christian and Esben Sloth Andersen, *Economic interdependence and innovative activity: an input-output analysis*, Cheltenham, UK: Edward Elgar, 1996.

Dosi, George, *Innovation, Organization and Economic Dynamics*, New York: Edward Elgar, 2001.

Eliasson, G., "Schumpeterian innovation, market structure, and the stability of industrial development," in H. Hanusch, ed., *Applications of Schumpeter's Ideas*, New York: Cambridge University Press, 1988.

Englmann, Frank C, "A Schumpeterian model of endogenous innovation and growth," *Journal of Evolutionary Economics*, 1994, *4* (3), 227–241.

Fagenberg, John, "International competitiveness," *Economic Journal*, 1988, *98*, 355–374.

Fölster, Stefan and Georgi Trofimov, "Industry evolution and R&D externalities," *journal of Economic Dynamics and Control*, 1997, *21* (10), 1727–1746.

Forum, World Economic, *Global Competitiveness Report*, New York: World Economic Forum, 2004.

Franke, Reiner, "A Reconsideration of Professor Iwai's Schumpeterian Dynamics," in Gallegati M. Gatti, D. and A. Kirman, eds., *Interaction and Market Structure*, New York: Springer, 2000, pp. 89–131.

Freeman, Christopher, *The economics of industrial innovation*, 2nd ed., London: Pinter Publishers, 1982.

Gallouj, Faïz, *Innovation in the service economy: the new wealth of nations*, Cheltenham, UK: Edward Elgar, 2002.

Gallouj, Faïz and Maria Savona, "Innovation in services: a review of the debate and a research agenda," *Journal of evolutionary economics*, 2008, *10* (2), 1–17.

Goodwin, Richard M, "A growth cycle," in C. Feinstein, ed., *Socialism, capitalism and economic growth*, Cambridge: Cambridge University Press, 1967, pp. 54–58.

Gordon, Robert, *The Measurement of Durable Good Prices*, Chicago: University of Chicago press, 1990.

Heertje, Arnold, "Schumpeter and technology change," in H. Hanusch, ed., *Application of Schumpeter's ideas*, New York: Cambridge University Press, 1988.

Iwai, Katsuhito, "Schumpeterian dynamics: An evolutionary model of innovation and imitation," *Journal of Economic Behavior & Organization*, 1984, *5* (2), 159–190.

Jacobides, Michael G and Sidney G Winter, "The co-evolution of capabilities and transaction costs: Explaining the institutional structure of production," *Strategic Management Journal*, 2005, *26* (5), 395–413.

Jorgenson, Dale W, Kevin J Stiroh, Robert J Gordon, and Daniel E Sichel, "Raising the speed limit: US economic growth in the information age," *Brookings papers on economic activity*, 2000, *2000* (1), 125–235.

Jovanovic, B, "Selection and the evolution of industry," *Econometrica*, 1982, *50*, 649–670.

Jr, Robert E Lucas, "Making a miracle," *Econometrica: Journal of the Econometric Society*, 1993, pp. 251–272.

Kelm, Matthias, "Schumpeter's theory of economic evolution: a Darwinian interpretation," *Journal of Evolutionary Economics*, 1997, *7* (2), 97–130.

Kim, Jong-Il, "Information technology and firm performance in Korea," in Takatoshi Itō, Andrew Rose et al., eds., *Growth and productivity in East Asia*, Chicago: University of Chicago Press, 2002.

Klein, Burton H, "Luck, necessity, and dynamic flexibility," 1988.

Laffont, Jean Jacques and Jean Tirole, *Competition in Telecommunications*, Cambridge, MA: The MIT Press, 2000.

Lansburg, Martin and David G Mayes, *Sources of productivity growth*, Vol. 49, Cambridge: Cambridge University Press, 1996.

Li, Wei and Lixin Colin Xu, "The impact of privatization and competition in the telecommunications sector around the world," *Journal of Law and Economics*, 2004, *47*, 75–98.

Lopez-Claros, Augusto and Yasmina N Mata, "Policies and Institutions Underpinning Country Innovation: Results from the Innovation Capacity Index," *The Innovation for Development Report*, 2010.

Malerba, Fredrick, *The semi conductor business*, London: Pinter Publishers, 1985.

Mario, J.L. Gaffard Amendola and Patrick Musso, "Innovative choice and competition process," in J. Kraft, ed., *The Process of Competition*, Cheltenham, UK: Edward Elgar, 2000.

Mazzucato, Mariana, *Firm Size, Innovation, and Market Structure: The Evolution of Industry Concentration and Instability*, Cheltenham: Edward Elgar Publishing, 2000.

Metcalfe, John, "Competition, evolution and the capital market," *Metroeconomica*, 1998, *4*, 127–154.

Metcalfe, J Stanley, *Evolutionary economics and creative destruction*, Vol. 1, London: Routledge, 1998.

Montobbio, Fabio, "An evolutionary model of industrial growth and structural change," *Structural Change and Economic Dynamics*, 2002, *13* (4), 387–414.

Nachum, Lilach, "International Business in a World of Increasing Returns," Technical Report, Cambridge, UK: University of Cambridge 2002.

Nelson, Richard R and Sidney G Winter, *An evolutionary theory of economic change*, Cambridge, Massachusetts: Belknap Press, 1982.

Norsworthy, John R and Show Ling Jang, *Empirical measurement and analysis of productivity and technological change: applications in high-technology and service industries*, North-Holland Amsterdam and New York, 1992.

Palokangas, Tapio, "Employment Cycles in a Growth Model with Creative Destruction," 2007, p. 423.

Pavitt, Keith, "Sectoral patterns of technical change: towards a taxonomy and a theory," *Research policy*, 1984, *13* (6), 343–373.

Phlips, Louis, *Competition Policy*, New York: Cambridge University Press, 1995.

Porter, Michael E, *Competitive advantage of nations*, New York: Free press, 1990.

Porter, Michael E, "Prosperity at Risk," *Cambridge, MA: Harvard University Press*, 2012.

Porter, Michael E and Scott Stern, "The Global Competitiveness Report," 2004.

Porter, M, *Global Competitiveness Report*, New York: World Economic Forum, 2004.

Qian, Xiaolei and Russell Smyth, "Growth accounting for the Chinese provinces 1990–2000: incorporating human capital accumulation," *Journal of Chinese economic and business studies*, 2006, *4* (1), 21–37.

Raynor, John and Ahmed, Yusuf, "Innovation and economic growth," *De Economist*, 2013, *145*, 12,12r–133.

Richardson, George Barclay, *Information and investment: a study in the working of the competitive economy*, Oxford: Clarendon Press, 1997.

Romer, Paul M, "Endogenous Technological Change," *Journal of Political Economy*, 1990, pp. 71–102.

Schumpeter, Joseph A, *Capitalism, socialism, and democracy*, London: Routledge, 1942.

Selten, Robert, "A simple model of Imperfect competition," *International Journal of Game Theory*, 1973, *2*, 141–201.

Sengupta, Jati, *Technology, innovations and growth*, New York: Palgrave Macmillan, 2011.

Sengupta, Jati, *Dynamics of industry growth*, New York: Springer, 2012.

Sengupta, Jati, *Technology Innovations and Growth*, New York: Palgrave Macmillan, 2010.

Sengupta, Jati, *Dynamics of Entry and Market Evolution*, New York: Palgrave Macmillan, 2007.

Sengupta, Jati and Phillip Fanchon, *Efficiency, Market Dynamics and Industry Growth*, New York: Palgrave Macmillan, 2009.

Spence, Michael, "Cost reduction, competition, and industry performance," *Econometrica*, 1984, 52:101.

Sutton, John, *Sunk costs and market structure: Price competition, advertising, and the evolution of concentration*, Cambridge, MA: The MIT press, 1991.

Sutton, John, *Technology and market structure: theory and history*, Cambridge, MA: MIT press, 1998.

Teece, David J, "Profiting from technological innovation: Implications for integration, collaboration, licensing and public policy," *Research policy*, 1986, *15* (6), 285–305.

Thompson, Peter, "Technological opportunity and the growth of knowledge: A Schumpeterian approach to measurement," *Journal of Evolutionary Economics*, 1996, *6* (1), 77–97.

Van Bael I and Bellis, *John Competition Law of the EEC*, London: CCH Ltd, 1990.

Williamson, Oliver E, *The Economic Institutions of Capitalism Firms Markets Relational Contracting*, New York: Free Press, 1985.

World Economic Forum, *Global Competitiveness Report*, New York, 2006.

Wu, Yanrui, *Productivity, efficiency and economic growth in China*, New York: Palgrave Macmillan, 2008.

MIX
Papier aus verantwortungsvollen Quellen
Paper from responsible sources
FSC® C105338

If you have any concerns about our products,
you can contact us on
ProductSafety@springernature.com

In case Publisher is established outside the EU,
the EU authorized representative is:
Springer Nature Customer Service Center GmbH
Europaplatz 3, 69115 Heidelberg, Germany

Printed by Libri Plureos GmbH
in Hamburg, Germany

Printed by Printforce, the Netherlands